Trout Theory

Walter Grayson

Published by Kokosnuss Press, 2025.

TROUT THEORY

First edition. February 19, 2025.

ISBN: 979-8230156147

Written by Walter Grayson.

Table of Contents

The River of Time

Time has long been one of humanity's most tantalizing mysteries. Across cultures and eras, we have sought to understand its nature, its flow, and our place within it. One of the most evocative metaphors for time is that of a flowing river—a ceaseless current carrying along everything in its path. We will explore the concept of time as a flowing river and introduce the idea of consciousness as a trout, navigating this ever-moving stream. We will trace the origins of this metaphor, relate it to classical theories of time travel, and then delve into how modern physics—through notions such as the block universe and quantum mechanics—offers us new ways of conceiving the flow of time.

The River as a Metaphor for Time

The image of a river has served as a powerful symbol for time throughout human history. In many ancient cultures, the river was seen as a boundary between worlds or as a path to the afterlife. The Greek philosopher Heraclitus famously observed, "No man ever steps in the same river twice," emphasizing the perpetual change inherent in life and nature. This ancient observation encapsulates the essence of time: it is not static, but dynamic and ever-shifting.

Just as a river continuously flows from its source to its mouth, time moves relentlessly forward. There is a natural progression—a current—that carries every moment into the next. Unlike a still pond, a river is never at rest; its waters are always in motion, shaping and reshaping the landscape as they move. This ceaseless movement offers an apt analogy for the progression of events in our lives and in the universe at large.

In literature and art, the river has frequently been used to symbolize the passage of time and the inevitability of change. Poets have likened the flow of time to the drifting of water, emphasizing both its beauty and its inexorable nature. Yet, within this flowing river, there exists the possibility of encountering eddies, whirlpools, or even moments when the current seems to recede—a notion that invites us to consider the possibility of deviating from the usual flow.

Consciousness as the Trout

Imagine, then, that within this vast river of time, there swims a single trout. This trout represents an individual's consciousness, a spark of awareness that navigates the currents of time. Just as a trout can swim with the current or, at times, make a determined effort to move against it, our consciousness might similarly experience time in ways that do not always align with the straightforward, unidirectional flow that we typically assume.

The trout metaphor opens up intriguing possibilities. In many time travel theories—especially those that focus on mental or consciousness-based time travel rather than the physical displacement of matter—the idea is that while our bodies remain anchored in the present, our minds might have the capacity to move upstream or downstream. In this view, the past, present, and future are not fixed states that we pass through like a series of train stations; rather, they are like segments of a river that can be revisited, reinterpreted, or even altered by the flow of consciousness.

This notion challenges our everyday experience of time. We usually perceive time as linear—a one-way march from birth to death. But what if our inner experience, our consciousness, is more like that of a trout exploring every eddy and current within the river? The trout may occasionally swim upstream, revisiting a moment from the past with the benefit of future knowledge, or drift downstream, letting go of a memory and embracing a new present. Such imagery provides a rich conceptual framework for discussing phenomena like déjà vu, the subjective acceleration or deceleration of time, and even the speculative possibility of "mental time travel."

Historical and Cultural Origins of the Metaphor

The idea of time as a river is not new. It is deeply embedded in mythology, philosophy, and literature. In many ancient cultures, rivers were seen as divine entities or pathways that connected different realms. The Nile, the Ganges, and the Euphrates, for example, were revered not only for their life-giving waters but also for their symbolic associations with the flow of time and the cycle of life and death.

In Eastern philosophies, the river often serves as a metaphor for the continuous cycle of reincarnation. Life is seen as a series of streams that converge, diverge, and eventually merge into a larger cosmic flow. This idea parallels the modern concept of time as a flowing continuum, where each moment is connected to those that came before and those that will follow.

Western literature, too, has embraced the river as a symbol of time. From the romantic poems of the 19th century to modern science fiction narratives, the river has been employed to evoke feelings of nostalgia, inevitability, and transformation. The river's dual nature—its calm surface and its hidden, sometimes turbulent currents—mirrors the complexity of human experience and the layered nature of time itself.

The metaphor has also found a home in the realm of science fiction, where it is often used to illustrate ideas about time travel. By likening time to a river and consciousness to a trout, writers can explore scenarios in which characters are able to "swim" against time, revisit past events, or even alter the course of history. This imagery serves as a bridge between abstract scientific theories and the human desire to understand and manipulate the flow of time.

Classical Time Travel Theories and the Trout Analogy

Time travel has long been a staple of both scientific inquiry and science fiction. Classical time travel theories have typically focused on the idea of physical movement through time—of a person or object being transported from one temporal point to another. However, when we shift our focus to the consciousness-based model, the trout metaphor provides a compelling visual: the body remains anchored in one place, while the mind navigates the temporal currents much like a fish swimming in a stream.

One of the key classical theories in time travel is the **Novikov Self-Consistency Principle**. This principle posits that any actions taken by a time traveler were always part of history, meaning that paradoxes—such as the famous "grandfather paradox"—are inherently impossible because the timeline self-corrects to maintain consistency. In the trout theory framework, the self-correcting nature of the river's current can be seen as analogous to this principle. Even if a trout (or consciousness) attempts to alter its course by swimming upstream, the prevailing current (the flow of time) ultimately guides it back to its destined path, preserving the overall narrative of events.

Another classical concept is the idea of the **block universe**. In this view, time is not a flowing river at all, but rather a fixed, four-dimensional block where past, present, and future coexist simultaneously. Every moment in time is laid out like a landscape, and our perception of the passage of time is merely a consequence of our consciousness moving along this fixed path. While this view may seem at odds with the dynamic imagery of a flowing river, it can be reconciled with the trout metaphor if we imagine that the river is not a force that changes the landscape, but rather a medium through which the trout (our consciousness) experiences a predetermined set of events. The trout can still choose which part of the river to navigate, even if the river itself is fixed in its overall course.

Quantum mechanics, with its probabilistic nature, offers yet another perspective on time and the flow of events. Concepts such as **quantum superposition** and **entanglement** suggest that particles (and, by extension, events) can exist in multiple states simultaneously until observed. In a way, this mirrors the idea of a trout that can be in multiple parts of the river at once—a metaphor for how our consciousness might access multiple moments in time. Some speculative theories even propose that consciousness itself might have quantum properties, potentially allowing it to interact with time in non-classical ways. While these ideas remain highly controversial and largely theoretical, they provide fertile ground for reimagining the relationship between time, matter, and mind.

Time in Modern Physics: The Block Universe and Beyond

Modern physics has given us a number of frameworks for understanding time. The block universe model, which we touched upon earlier, treats time as another dimension similar to space. In this model, all moments in time exist simultaneously in a four-dimensional spacetime continuum. This notion challenges our everyday experience of time as something that flows or passes, suggesting instead that the past, present, and future are equally real and immutable.

In a block universe, the image of a river may seem less applicable at first glance. However, if we reinterpret the river not as the passage of time but as the experience of time by a conscious observer, the analogy holds strong. The trout—representing our consciousness—moves along a predetermined landscape of events, experiencing the flow of time subjectively even if, in an objective sense, time is fixed. This perspective helps bridge the gap between the impersonal, unchanging nature of the block universe and our personal, lived experience of time as something that is dynamic and ever-changing.

Quantum mechanics further complicates our understanding of time. Unlike the deterministic view of classical mechanics, quantum theory introduces elements of uncertainty and probability into the fabric of reality. The famous **Copenhagen interpretation** posits that particles exist in a state of superposition until measured, at which point they collapse into a definite state. This idea has led some physicists and philosophers to speculate that time, too, may not be as rigid as it appears.

Consider, for instance, the notion of **retrocausality**—the idea that future events can influence past events. While this concept remains highly speculative and controversial, it opens the door to the possibility that consciousness (or information) could travel "backward" along the river of time. In such a scenario, the trout might find itself swimming upstream, revisiting past events with the benefit of future knowledge. Although experimental evidence for retrocausality is scant, the very possibility challenges our conventional understanding of cause and effect, suggesting that the river of time might have hidden undercurrents that allow for backward flow under special circumstances.

Another fascinating development in modern physics is the study of **time dilation** in the context of Einstein's theory of relativity. According to relativity, time is not absolute—it can stretch or compress depending on an observer's velocity and the strength of gravitational fields. Astronauts traveling at high speeds or residing near massive objects experience time differently from those on Earth. This relativistic effect, while typically minute in everyday experience, illustrates that time is not a uniform river but rather a malleable medium, subject to the influences of motion and gravity. In our trout metaphor, this would be akin to certain stretches of the river where the current is altered—faster in some regions and slower in others—affecting the trout's journey in unexpected ways.

Intersecting the Metaphor with Human Experience

Beyond the realms of abstract physics and high theory, the river metaphor resonates deeply with human experience. Every individual has moments when time seems to fly by and moments when it drags endlessly. Our memories often feel like snapshots taken along the banks of a river, glimpses of water that continue to flow past even as we try to capture them. In these moments, our consciousness acts much like a trout that occasionally stops to take in its surroundings, only to be swept along by the relentless current once more.

Moreover, the idea of mental time travel—of revisiting past memories or envisioning future possibilities—has been explored extensively in psychology and neuroscience. When we recall a cherished memory or anticipate an important future event, it is as though our consciousness momentarily leaves the present, swimming upstream or downstream to visit another part of the river. Such experiences underscore the fluidity of time in our subjective perception, suggesting that while the river of time may be inexorable on a grand scale, our personal journeys within it are rich with detours, reflections, and even the occasional loop.

The interplay between memory, perception, and time has profound implications for our understanding of identity and personal growth. Just as a trout might navigate through different parts of the river, our consciousness weaves together past experiences, present awareness, and future aspirations into a coherent narrative. This narrative, in turn, shapes how we perceive ourselves and our place in the world. It is a reminder that while time itself might be an unyielding current, our inner experiences give that current meaning, depth, and context.

A Synthesis: The Dynamic Dance of Time and Consciousness

In synthesizing these ideas, the river of time becomes more than just a metaphor—it evolves into a dynamic framework for understanding both the physical and experiential aspects of time. On one level, physics tells us that time is a dimension woven into the fabric of the universe, subject to the laws of relativity and quantum mechanics. On another level, our subjective experience of time is fluid, mutable, and intimately tied to our consciousness.

The trout, representing our mind or awareness, is not merely a passive observer adrift in the current. Instead, it is an active participant in the dance of time—sometimes swimming with the flow, sometimes daring to venture upstream in moments of reflection, and occasionally drifting downstream as memories fade and new experiences emerge. This active engagement with time invites us to reconsider long-held assumptions about fate, destiny, and the very nature of free will. If our consciousness can, even in theory, traverse different segments of the river, what does that say about our ability to change the course of our lives? Could moments of introspection or sudden insights be understood as brief excursions against the current, moments when the trout momentarily escapes the inexorable pull of the river?

While current scientific theories do not yet provide a definitive mechanism for such mental time travel, the metaphor serves as a powerful tool for imagining the possibilities. It encourages us to view time not as a rigid, unalterable sequence of events but as a dynamic, interactive process—one in which our minds have the potential to navigate, interpret, and perhaps even influence the currents that shape our lives.

Looking Forward: Questions and Possibilities

As we conclude this exploration of the river of time and the consciousness that swims within it, we are left with as many questions as answers. What does it truly mean to travel through time if our journey is guided not by physical movement but by the flow of our inner experiences? Can moments of profound insight or sudden recollection be seen as evidence of a deeper connection between consciousness and the fabric of time? And, more provocatively, if our minds are capable of navigating this river in unconventional ways, what might that imply for the future of time travel—both as a scientific pursuit and as a narrative device in literature and film?

These questions invite us to look beyond the established boundaries of physics and psychology, to imagine a future where the lines between past, present, and future are more fluid than we currently believe. They challenge us to rethink our assumptions about causality, destiny, and the nature of change. Whether through the lens of advanced scientific theory or the evocative power of metaphor, the river of time remains a potent symbol—one that continues to inspire both wonder and inquiry.

In the coming chapters, we will dig deeper into the mechanics of consciousness time travel, explore how these ideas have been represented in popular culture, and examine the philosophical and ethical implications of reinterpreting our relationship with time. For now, the image of a lone trout navigating a vast, flowing river stands as a reminder of the profound mystery of time—a mystery that is as much about the universe as it is about the inner workings of our own minds.

The journey through the river of time is both an exploration of scientific theory and an invitation to reflect on the nature of our own existence. By considering time as a flowing river and our consciousness as a trout navigating its currents, we gain a new perspective on familiar experiences—one that highlights the fluidity of memory, the possibility of revisiting the past, and the ever-present influence of the future on our current reality.

This metaphor, rich in cultural, historical, and scientific significance, challenges us to think about time not as a linear progression but as a dynamic and multifaceted process.

Whether we view the river as a fixed landscape of events in a block universe or as a malleable medium subject to the quirks of quantum mechanics and relativity, the image of the trout offers a compelling vision of human consciousness—a force capable of probing the depths of time, sometimes resisting the current, and always striving to understand the complex dance between memory, experience, and destiny.

As we continue our exploration, keep in mind that this river is not just an abstract concept. It is a living, breathing symbol of our ongoing quest to make sense of the world around us—a quest that unites the realms of science, art, and philosophy in a timeless dialogue about what it means to exist. The trout in the river of time is both a wanderer and a participant in the grand tapestry of existence, a reminder that even as the current carries us inexorably forward, we possess the capacity to reflect, to dream, and perhaps even to momentarily swim against the tide.

With this understanding, we open our minds to the possibility that time, far from being a rigid and unyielding force, is a dynamic medium in which our consciousness plays an active and creative role. The journey has only just begun, and as we traverse these conceptual currents, we may yet discover that the secrets of the universe are not hidden in the stars alone, but also within the depths of our own awareness.

The Mind as the Time Machine

Time travel is a concept that has long captivated the human imagination, but what if the true journey through time does not involve a physical leap from one era to another, but instead is undertaken by the mind itself? We explore the fascinating idea that while physical time travel remains confined to the realms of science fiction and theoretical physics, the human mind may already be an active traveler through time. We will delve into the distinctions between physical and mental time travel, investigate theories on memory and déjà vu, and consider whether the brain might "reboot" to an earlier state. Finally, we will examine how modern neuroscience and quantum consciousness theories—such as the Penrose-Hameroff Orchestrated Objective Reduction (Orch-OR) hypothesis—could provide insights into the notion that our consciousness might serve as its own time machine.

Physical Versus Mental Time Travel

Defining the Two Modes of Time Travel

Physical time travel typically evokes images of machines and devices transporting bodies through the fabric of spacetime—whether by warping gravity, harnessing wormholes, or bending the rules of relativity. This form of time travel implies that an entire physical entity, along with its biological and material components, is relocated from one point in time to another. Despite its allure, such physical time travel remains a theoretical construct in physics, laden with paradoxes and formidable technological challenges.

In contrast, mental time travel does not require the body to leave its current temporal position. Instead, it involves the ability of the human mind to revisit the past, anticipate the future, or even reframe present experiences through memory, imagination, and introspection. This type of travel occurs entirely within the neural circuits and cognitive processes of the brain. Here, time is experienced subjectively rather than objectively—a phenomenon that allows the mind to "travel" without the need for any physical displacement.

The Practicality of Mental Journeys

Consider the experience of recalling a long-forgotten childhood memory. As you close your eyes, you might vividly re-experience the sights, sounds, and emotions of that past moment. In this way, your consciousness momentarily departs from the present, diving into the depths of your personal history. Unlike the fixed timeline of physical time travel, mental time travel is fluid, intimate, and available to anyone who dares to reflect on their past.

Similarly, when you plan for an upcoming event or fantasize about possible future outcomes, you are engaging in a form of prospective mental time travel. This cognitive ability to simulate future scenarios—often called "episodic future thinking"—is a crucial component of decision-making and problem-solving. By mentally projecting ourselves into possible futures, we can anticipate consequences, rehearse actions, and refine our strategies for navigating an uncertain world.

Bridging the Divide: The Brain as the Portal

Where physical time travel grapples with the grand scales of cosmology and the immutable laws of physics, mental time travel taps into the dynamic and subjective experience of being human. The brain, with its vast network of neurons and synaptic connections, functions as a portal through which we experience the continuum of time. It is here that past, present, and future intermingle, allowing us to traverse memories and imaginations in ways that defy the conventional constraints of chronological order.

The distinction is crucial: while physical time travel might allow for paradoxical encounters with one's past self or the alteration of historical events, mental time travel is inherently personal and reflective. It does not alter external reality but rather reshapes our internal narrative. The brain's capacity to store and retrieve information across time enables us to re-experience and reinterpret moments that have long passed. In doing so, it provides a natural mechanism for learning, healing, and even reinventing ourselves.

Theories on Memory, Déjà Vu, and the Brain's Temporal Navigation

Memory: The Archive of Personal Time

Memory is often described as the mind's personal archive—a repository of experiences, emotions, and knowledge.

Neuroscientific research has revealed that memory is not a static record but a dynamic process, continuously reconstructed and reinterpreted with each recollection. This malleability is at the heart of mental time travel. Each time we remember an event, we are not simply replaying a stored video; we are actively reassembling a past experience, often infusing it with our current emotional state, perspective, and context.

The hippocampus, a seahorse-shaped structure nestled within the brain, plays a pivotal role in this process. It is instrumental in consolidating short-term memories into long-term storage and in reactivating these memories when needed. This neural mechanism allows us to "relive" the past and use it as a guide for future behavior. The act of remembering, therefore, is not merely a passive retrieval of data but an active reconstruction that can shape our understanding of who we are.

Déjà Vu: A Glimpse of Temporal Overlap?

One of the most intriguing phenomena related to mental time travel is déjà vu—a sensation that one has already experienced a current situation, even when logically, it seems impossible. Scientists have long debated the origins of déjà vu, proposing various theories that range from neural misfires to the possibility of parallel processing within the brain.

Some researchers suggest that déjà vu may occur when the brain processes sensory information in duplicate channels, causing a temporary overlap between perception and memory. In this view, the brain might inadvertently "misfile" current experiences as memories, creating the eerie impression that the moment has occurred before. This phenomenon could be interpreted as a fleeting instance of mental time travel, where the boundary between the present and the past becomes blurred.

Another perspective considers déjà vu as evidence of the brain's capacity to momentarily access stored memories that are similar but not identical to the current context. The familiarity experienced during déjà vu may arise from subconscious recognition of environmental cues that echo past experiences. Although the sensation is fleeting and often puzzling, it serves as a reminder that our internal timeline is not a rigid sequence but a complex tapestry of overlapping moments and associations.

Can the Brain "Reboot" to an Earlier State?

The notion of a brain "reboot"—a process whereby the mind can revert to an earlier state—raises profound questions about identity, continuity, and the nature of consciousness. Some theories posit that under certain conditions, such as during deep meditation, near-death experiences, or intense emotional episodes, the brain might temporarily reset its pattern of activity, allowing for an altered perception of time.

In such moments, individuals often report feeling detached from the present, as if they are reliving a past event or experiencing a timeless state. This subjective experience of time suspension or reversal is akin to a mental reset, where the usual flow of time seems to halt or even reverse. Researchers have noted that such states are sometimes accompanied by distinct neural patterns, including changes in brainwave activity that resemble those observed during sleep or trance-like conditions.

The idea of a neural "reboot" is further supported by observations in clinical settings. For instance, patients with certain neurological conditions, such as temporal lobe epilepsy, sometimes report vivid flashbacks or a sense of reliving past experiences during seizures. These episodes, though pathological in nature, hint at the possibility that the brain retains the capacity to re-access earlier states of consciousness under specific circumstances.

While the concept of a full-scale reboot—where the brain entirely reverts to a previous state—remains speculative, the evidence suggests that our neural circuits are more flexible than previously thought. This flexibility might allow for temporary excursions into our past, providing the cognitive basis for what we might call mental time travel.

Neuroscience and the Quantum Consciousness Perspective

The Neural Substrate of Temporal Experience

To understand how the brain might function as a time machine, it is essential to examine the neural substrates that underpin our experience of time. Recent advances in neuroscience have identified several key brain regions involved in temporal processing. The prefrontal cortex, for example, is crucial for planning and decision-making, enabling us to simulate future scenarios. Meanwhile, the medial temporal lobe, which includes the hippocampus, is central to encoding and retrieving memories.

Neurons in these regions communicate through complex networks of synapses, transmitting signals that encode the passage of time. Some studies have proposed that certain neurons, often referred to as "time cells," are capable of firing in patterns that correlate with the timing of events. These cells could serve as an internal clock, helping to organize memories and experiences along a temporal axis. The existence of such cells supports the notion that the brain is intrinsically designed to process time—not only in a linear sequence but also in a way that allows for the flexible manipulation of past, present, and future information.

Quantum Consciousness and the Orch-OR Hypothesis

Among the more radical theories that attempt to bridge neuroscience with the mysteries of time is the Penrose-Hameroff Orch-OR (Orchestrated Objective Reduction) hypothesis. Proposed by mathematician and physicist Sir Roger Penrose and anesthesiologist Stuart Hameroff, Orch-OR suggests that consciousness arises from quantum processes occurring within microtubules—tiny, tubular structures that form part of the cytoskeleton in neurons.

According to Orch-OR, these quantum processes are capable of supporting non-computational states that contribute to conscious awareness. In this framework, the brain is not merely a classical information processor but also a quantum system that may tap into the probabilistic nature of the universe. If true, this could have profound implications for our understanding of time and consciousness.

Quantum mechanics introduces the concept of superposition, where particles exist in multiple states simultaneously until measured. When applied to consciousness, this suggests that the brain might, in some fashion, access a range of potential outcomes or temporal states before collapsing into a single experienced reality. In other words, just as quantum particles can be in several places at once, our consciousness might be capable of spanning different moments in time—albeit briefly and under highly specific conditions.

Implications of Quantum Processes for Mental Time Travel

The Orch-OR theory is not without controversy. Critics argue that the warm, wet environment of the brain is inhospitable to the delicate quantum effects observed in controlled laboratory settings. Nonetheless, proponents of the theory point to evidence that some biological systems—such as photosynthetic complexes and avian navigation mechanisms—do appear to exploit quantum phenomena. If the brain is similarly capable of harnessing quantum effects, then mental time travel may be more than just a metaphor; it could be an emergent property of quantum processes interacting with neural networks.

Under the Orch-OR model, the brain's capacity to process time might be enhanced by quantum coherence and entanglement. Quantum coherence allows particles to remain in sync over distances, while entanglement links particles such that the state of one immediately influences the state of another, regardless of distance. Translated into neural terms, these phenomena could provide a mechanism by which disparate regions of the brain—responsible for memory, emotion, and sensory processing—communicate in a unified, temporally dynamic fashion. This coordinated activity might enable the kind of mental excursions that we interpret as revisiting the past or anticipating the future.

Moreover, the probabilistic nature of quantum mechanics may offer an explanation for the fluidity of memory and perception. Since quantum states are inherently uncertain until observed, the mind's recollection of events could be seen as a probabilistic reconstruction rather than a fixed playback. This could account for the malleability of memory—the way in which details shift over time or become imbued with the emotions of the present moment. Such a mechanism aligns well with the idea of mental time travel, where the boundaries between distinct temporal states are blurred and reassembled according to the needs of the individual.

The Interplay of Memory, Consciousness, and Temporal Perception

Memory as a Living Process

In exploring mental time travel, it is essential to recognize that memory is not a static repository of data but an active, living process. Every act of recollection involves the reactivation of neural circuits that were originally engaged when the memory was formed. This reactivation is not a perfect reproduction but a creative reconstruction—a reinterpretation that integrates both the original sensory details and the current mental state of the individual.

This dynamic nature of memory means that every time we "travel" back to a moment in our past, we do so in a way that is uniquely influenced by our present circumstances. A childhood memory might be re-experienced with the wisdom of adulthood, or a painful event might be revisited with the healing insights of time. In this sense, the mind acts as a time machine by continually reshaping the narrative of our lives, merging the past with the present to create an evolving sense of identity.

Temporal Illusions and the Flexibility of the Mind

Beyond deliberate recollection, the mind is prone to temporal illusions—phenomena that distort our perception of time. In addition to déjà vu, other effects such as time dilation (where time appears to slow down during moments of high stress or excitement) and the subjective compression of time during repetitive tasks illustrate how malleable our experience of time can be.

These temporal illusions reveal that the brain does not function as a passive recorder of events, but as an active interpreter of temporal information. In doing so, it allows for moments where the boundaries between past, present, and future become indistinct. Such experiences, though often fleeting, offer tantalizing glimpses into the possibility that mental time travel is not only real but is an intrinsic part of human cognition.

Integrating Neuroscience and Subjective Experience

The intersection of neuroscience with the subjective experience of time provides a rich landscape for understanding mental time travel. Neuroimaging studies have shown that when individuals recall vivid autobiographical memories, a network of brain regions—including the hippocampus, prefrontal cortex, and posterior cingulate cortex—becomes activated. This network is similarly engaged when subjects imagine future scenarios, suggesting that the same neural machinery underlies both recollection and anticipation.

This overlap supports the view that our mental exploration of time is not divided into separate modules for past and future but is rather a unified process. The brain appears to navigate a continuum where memories and imaginings coexist, blurring the strict lines of temporal order. In this way, mental time travel is a natural function of our cognitive architecture—a process that not only allows us to reflect on who we have been but also to envision who we might become.

Challenges, Criticisms, and Future Directions

Skepticism Toward Quantum Consciousness

While the Orch-OR hypothesis and other quantum consciousness theories offer an intriguing framework for understanding mental time travel, they remain highly controversial. Many neuroscientists argue that the brain's environment is too chaotic for quantum coherence to play a significant role in cognition. Critics point out that most quantum effects require extremely low temperatures and controlled conditions, which are not present in the human brain.

Despite these criticisms, research in quantum biology has revealed that some biological processes do, in fact, exploit quantum phenomena. The debate continues, and while definitive proof of quantum consciousness remains elusive, the possibility invites further inquiry into how the principles of quantum mechanics might intersect with neurobiology. As experimental techniques advance, we may find that the brain is capable of subtle quantum effects that contribute to its extraordinary flexibility in processing time.

The Limitations of Memory as a Time Machine

Even if we accept that the brain functions as a kind of time machine through its handling of memory and imagination, there are inherent limitations. Memories are notoriously fallible—they are subject to distortion, decay, and even the influence of suggestion. This imperfection means that our mental journeys through time are not accurate reconstructions of the past, but rather personal narratives shaped by both objective events and subjective interpretation.

Moreover, the brain's tendency to fill in gaps or alter details over time can lead to the creation of "false memories," further complicating our understanding of mental time travel. These limitations do not diminish the profound role that memory plays in our experience of time; rather, they underscore the need for a nuanced view that recognizes the interplay between objective reality and subjective perception.

Future Research and the Promise of New Technologies

Looking ahead, advances in neuroscience and cognitive science may open new avenues for exploring mental time travel. Techniques such as high-resolution brain imaging, optogenetics, and neural decoding are beginning to reveal the intricate ways in which the brain stores and retrieves temporal information. These tools may one day allow us to "map" the neural pathways involved in mental time travel, offering insights into how we can harness or enhance this natural capability.

Additionally, as research into quantum biology progresses, scientists may uncover new evidence supporting the role of quantum effects in neural processes. Should these findings materialize, they could fundamentally alter our understanding of consciousness and its relationship to time. In such a scenario, mental time travel would not merely be a metaphor but a tangible aspect of the human experience—one that could lead to novel therapeutic approaches for memory disorders, mental health challenges, and even cognitive enhancement.

Embracing the Mind's Temporal Odyssey

The concept of the mind as a time machine invites us to reimagine our relationship with time. While physical time travel remains in the realm of theoretical physics and speculative fiction, our internal journeys through memory, imagination, and consciousness reveal a rich tapestry of temporal exploration. The brain, with its remarkable capacity to store, reconstruct, and simulate experiences, serves as both a repository of our past and a laboratory for our future.

By understanding the neural mechanisms that underpin memory and temporal perception, we gain insight into how the mind can traverse the continuum of time. Whether it is through the vivid recollection of a long-forgotten event, the uncanny familiarity of déjà vu, or the imaginative projection into future possibilities, mental time travel is an intrinsic part of the human experience.

Furthermore, emerging theories in quantum consciousness—despite their controversies—offer provocative possibilities for how the brain might operate at the intersection of quantum mechanics and neural computation. While challenges remain in reconciling these ideas with established neuroscientific principles, they nonetheless open the door to innovative ways of thinking about consciousness, memory, and time.

In embracing the notion of the mind as a time machine, we are invited to appreciate the complexity and wonder of our own cognitive processes. Our ability to traverse time mentally is not just a quirk of memory but a testament to the dynamic, ever-evolving nature of our consciousness. It is a reminder that, even as we move inexorably forward in the present, we carry within us the echoes of the past and the seeds of the future.

As we continue our exploration of time and consciousness in subsequent chapters, the insights gleaned from mental time travel will serve as a foundation for understanding how our internal experiences shape not only our personal identities but also our perceptions of reality itself. The brain's capacity to journey through time—however fleeting or subjective—remains one of the most profound aspects of human existence, urging us to question what it truly means to live, remember, and imagine.

In this ever-unfolding odyssey, the mind stands as both traveler and guide, a time machine capable of unlocking the mysteries of our temporal landscape. Whether through the reactivation of cherished memories or the construction of hopeful visions for the future, our inner journeys enrich our lives and deepen our understanding of the nature of time. The exploration of mental time travel thus becomes a celebration of the human spirit—its resilience, creativity, and unyielding quest for meaning in the vast continuum of existence.

Loops, Paradoxes, and the Self-Correcting Timeline

Time travel—whether physical or mental—inevitably raises the specter of paradoxes and logical conundrums. Let's dig into the labyrinth of loops, paradoxes, and the possibility of a self-correcting timeline, exploring how consciousness-based time travel might confront these challenges. We will examine the famous Novikov self-consistency principle and contrast it with the idea of chaotic, unpredictable changes in history. Furthermore, we will consider how history might resist certain alterations while allowing others, and what this implies for our understanding of causality and free will.

The Nature of Temporal Loops and Paradoxes

Understanding Temporal Loops

Temporal loops represent a class of time travel scenarios in which events repeat in a cyclic fashion. Imagine a scenario where a time traveler's actions cause events to unfold in a loop, such that their journey becomes an integral part of the historical timeline. In such a case, the loop is self-contained: the traveler's future actions ensure the very conditions that allowed them to time travel in the first place.

A classic example of a temporal loop is the "predestination paradox," where the time traveler's intervention in the past is exactly what leads to the circumstances that cause the time travel event. For instance, suppose a scientist sends a message back in time, and that very message inspires their younger self to become a scientist. Here, the loop is closed—without the future intervention, the chain of events leading to the time traveler's existence might never occur.

In the realm of consciousness-based time travel, loops may not require physical relocation through time but instead can manifest as cycles in mental states. A person might repeatedly relive or re-experience critical moments in their life, each iteration reinforcing the narrative of their past. These mental loops could be considered analogous to the physical loops depicted in many time travel stories, with the brain's memory circuits forming a self-sustaining cycle of recollection and reinterpretation.

The Grandfather Paradox and Its Variants

Perhaps the most famous time travel paradox is the grandfather paradox, which poses the following scenario: if a time traveler goes back in time and prevents their grandfather from meeting their grandmother, the traveler's own birth would be impossible, thereby negating the possibility of the time travel event itself. This contradiction raises the question: how can actions in the past affect a future that no longer exists?

Consciousness-based time travel introduces a subtler form of this paradox. Rather than physically altering the past, a person might attempt to change their internal narrative or reinterpret past events in a way that seems to "undo" established memories. For example, if someone were to experience a profound insight that recontextualizes a traumatic memory, does that not change the very fabric of their personal history? While such mental alterations may not cause a paradox in the physical sense, they raise questions about the stability of the self and the coherence of one's identity over time.

Other paradoxes emerge when we consider the possibility of multiple overlapping consciousness streams. Imagine if an individual's mind could split—each branch experiencing different outcomes based on divergent recollections or decisions. Would these branches converge into a single narrative, or would they remain forever separate? Such scenarios force us to grapple with the potential multiplicity of selfhood and the fluidity of identity, highlighting the complex interplay between memory, consciousness, and the passage of time.

The "Bootstrap" Paradox

A related conundrum is the bootstrap paradox. In this scenario, an object, piece of information, or even a state of mind appears to be "self-originating." Consider a situation where a person receives an idea from their future self and then acts on it, thereby ensuring that the idea exists in the first place. The idea is never created—it simply exists in a self-sustaining loop.

For consciousness-based time travel, the bootstrap paradox may occur when a transformative insight or epiphany is passed backward through one's own mental timeline. The brain could "borrow" from its future repository of knowledge, allowing an individual to solve a complex problem before they are logically capable of doing so. Such a scenario challenges our conventional notions of learning and self-development. If the mind can retrieve information from its own future iterations, then what is the origin of that knowledge? The bootstrap paradox pushes us to reconsider the boundaries between memory, intuition, and genuine foresight.

Novikov's Self-Consistency Principle

An Overview of the Principle

In the face of such paradoxes, one of the most influential proposals to preserve logical consistency in time travel scenarios is the Novikov self-consistency principle. Formulated by physicist Igor Novikov in the 1980s, this principle asserts that any event occurring in a timeline that includes time travel must be self-consistent. In other words, the actions of a time traveler cannot create a contradiction or alter history in a way that would prevent the time travel event from occurring.

According to Novikov's principle, even if a person's consciousness were to travel back in time, the sequence of events would adjust itself in such a way that any potential paradox is avoided. The timeline "self-corrects," ensuring that every action taken by the traveler is already part of history. This concept not only applies to physical time travel but can also be extended to mental time travel, where internal loops of memory and reinterpretation must eventually reconcile with the overarching narrative of the self.

Applying Self-Consistency to Consciousness-Based Time Travel

When we consider the mind as a time machine, the self-correcting nature of the timeline may manifest as the inherent stability of our personal narrative. Despite the malleability of memory and the potential for reinterpretation, our identity remains coherent over time. This suggests that there are constraints—perhaps at the neural or psychological level—that prevent our internal time travel from producing contradictions in our lived experience.

For instance, if an individual repeatedly revisits a formative memory with the intent of changing its emotional impact, the self-consistency principle would imply that the alteration is already integrated into the narrative of their life. No matter how many times they "re-experience" the event, the essential facts remain unaltered, and the overall trajectory of their identity is preserved. The mind might allow for variations in perception or emotional intensity, but the fundamental structure of the memory is self-correcting, ensuring that a paradoxical state—such as a memory that negates its own existence—never occurs.

Evidence from Cognitive Science and Psychology

Empirical studies in psychology provide indirect support for a self-correcting timeline in the realm of consciousness. Research on memory reconsolidation—the process by which recalled memories become labile and then restabilize—suggests that while memories can be updated or modified, they do so within a framework that maintains overall coherence. Even when a memory is altered by new information or reinterpretation, it does not lead to a breakdown in the continuity of the self.

Furthermore, cognitive dissonance theory, which examines the discomfort associated with holding contradictory beliefs or memories, indicates that the human mind is predisposed to resolve inconsistencies. This resolution process can be viewed as a form of internal self-correction, where the mind adjusts its narrative to maintain consistency and reduce psychological conflict. In this sense, our inner experience of time is safeguarded by mechanisms that ensure new insights or reinterpretations do not result in an irreconcilable paradox.

Chaotic Changes and the Possibility of Divergent Timelines

The Concept of Chaotic Temporal Dynamics

While the self-consistency principle offers a neat solution to many time travel paradoxes, not all theories agree that history must remain entirely unaltered. Some models propose that the timeline might be inherently chaotic, allowing for small alterations that, while not catastrophic, could accumulate over time. In such models, history is not an immovable monolith but rather a dynamic system sensitive to initial conditions.

In the context of consciousness-based time travel, chaotic changes could manifest as subtle shifts in how events are remembered or interpreted. A single act of mental time travel might introduce a minute variation in an individual's recollection, which in turn could influence future decisions and perceptions. Over time, these small deviations could lead to a cascade of changes—a butterfly effect of sorts within the inner workings of the mind.

The Butterfly Effect and Consciousness

The butterfly effect, a term popularized by chaos theory, posits that minor disturbances can lead to significant, unpredictable consequences over time. When applied to mental time travel, this concept suggests that even slight modifications in memory or perception could set off a chain reaction that alters an individual's future choices or even their personality.

For example, consider a scenario where a person, during a moment of reflection, decides to reinterpret a past failure as a learning experience rather than a source of shame. This reinterpretation might lead to increased confidence and a greater willingness to take risks, thereby altering their subsequent life trajectory. Although the change in perspective may seem minor at first, the cumulative effect over the years could be profound—resulting in a life that is markedly different from the one that would have unfolded without that mental shift.

However, the butterfly effect in the context of consciousness-based time travel is tempered by the fact that the human mind possesses mechanisms for maintaining a coherent narrative. While small changes may be possible, the overall structure of one's identity and life story tends to resist radical alterations. This interplay between chaotic influence and self-correction raises important questions: To what extent can our memories and internal narratives be changed without destabilizing our sense of self? And are there limits beyond which alterations become impossible because the timeline "pushes back" to preserve consistency?

The Role of Free Will and Determinism

The tension between self-correction and chaotic change brings us to the age-old debate between free will and determinism. If the timeline is self-correcting, then every attempt at change—whether mental or physical—is ultimately predetermined, and free will may be more illusion than reality. According to a strict interpretation of the Novikov principle, any decision made by a time traveler, even one based on foreknowledge or introspection, was always part of history. In this view, free will is subsumed by the necessity of a consistent timeline.

On the other hand, the possibility of chaotic changes within the mind suggests that there is room for genuine agency. An individual might be able to influence their internal narrative in ways that have meaningful, if subtle, effects on their future. In this perspective, free will exists not as a means to rewrite history in dramatic fashion, but as a mechanism for gradually reshaping one's path through a complex and responsive temporal landscape. The debate over free will and determinism is far from settled, yet the exploration of mental time travel offers a unique vantage point from which to consider these philosophical questions.

How History Might Resist or Allow Alterations

The Inertia of Historical Events

One perspective on time and causality posits that history, whether in the physical world or within the contours of human memory, possesses a kind of inertia. Major events and turning points in one's life—or in the collective history of a society—tend to be self-sustaining. Once a significant event has been established, it can be remarkably resistant to change. This inertia is partly due to the interconnected nature of events: a major decision or incident ripples outward, influencing numerous subsequent occurrences, thereby reinforcing its own place in the timeline.

From the viewpoint of consciousness-based time travel, the inertia of historical events may serve as a bulwark against paradoxical alterations. Even if a person's mind were to revisit a crucial moment, the broader context in which that event occurred may exert such a strong influence that any attempted reinterpretation is ultimately absorbed into the existing narrative. In other words, history "pushes back," ensuring that fundamental events remain unchanged despite individual efforts to view them differently. This self-correcting behavior underscores the robustness of our internal timeline—it is not easily malleable by isolated acts of recollection or reinterpretation.

The Threshold of Change: Minor Shifts vs. Major Alterations

While the inertia of history may prevent drastic rewrites of one's past, it is possible that minor shifts in perception or memory could occur without triggering catastrophic paradoxes. These subtle changes might involve re-evaluating the emotional significance of an event, discovering new insights about one's motivations, or even reordering the sequence in which memories are recalled. In such cases, the alterations do not so much rewrite history as they add nuance to it.

For instance, consider someone who has long harbored a negative view of a particular personal failure. Through mental time travel—perhaps in the form of deep reflection or guided meditation—they come to see that failure not as a final defeat but as an integral step in their growth. While the factual details of the event remain intact, the emotional impact and its role in shaping future decisions may be significantly altered. Here, the timeline is not overhauled; rather, it is refined. Such minor shifts might even be essential to personal development, allowing individuals to adapt their narratives without undoing the fabric of their lived experience.

Conversely, if one were to attempt a major alteration—such as erasing an entire period of memory or fundamentally reinterpreting a series of interconnected events—the timeline might resist these changes. In a self-correcting system, there could be inherent limits to how far the internal narrative can deviate from established history. Attempts at radical rewrites might be met with cognitive dissonance or even subconscious "corrections" that restore the original order of events. This resistance could be seen as the mind's way of preserving continuity, ensuring that its internal timeline remains stable despite the fluidity of memory and perception.

External Influences and Collective History

It is also important to consider how external factors and collective history interact with individual mental time travel.

Our personal narratives do not exist in isolation; they are part of a broader tapestry of shared experiences, cultural memory, and historical events. In this context, the self-correcting properties of time may extend beyond the individual to the societal level. Historical events that have shaped entire communities or nations may be even more resistant to change, as they are continuously reinforced by collective memory and documentation.

When an individual engages in mental time travel to reinterpret a personal event, that reinterpretation might eventually interact with the wider historical narrative. For example, a reappraisal of a traumatic event might not only alter personal memory but could also influence how that event is understood within a cultural context. However, the collective narrative is bolstered by multiple accounts, records, and communal experiences, which makes wholesale alterations unlikely. The interplay between personal memory and collective history illustrates that while individual minds may navigate temporal loops and paradoxes, the broader timeline retains a degree of rigidity—a safeguard against the chaos of unlimited alteration.

Reconciling Paradoxes: A Spectrum Between Self-Correction and Chaos

A Continuum of Temporal Dynamics

The discussion so far suggests that the nature of time travel—whether physical or mental—may lie on a continuum between absolute self-correction and unpredictable chaos. At one end of the spectrum, the Novikov self-consistency principle demands that all events, including those instigated by time travelers, are predetermined and logically consistent. At the other end, chaotic models of time posit that even slight variations might snowball into significant changes, allowing for a degree of unpredictability in the unfolding of history.

In the context of consciousness-based time travel, the interplay between these extremes can be thought of as a delicate balance. The mind's ability to reinterpret memories and imagine alternative futures might introduce small variations—micro-level chaotic changes—that enrich the internal narrative without undermining the overall structure of one's life. These adjustments, rather than being paradoxical, could be seen as the natural evolution of a living, breathing timeline. In contrast, any attempt to instigate a fundamental rewrite of history would likely trigger a self-correcting response, as the underlying inertia of established events reasserts itself.

The Role of Constraints and Feedback Mechanisms

One way to conceptualize the balance between chaos and self-correction is to imagine the timeline as a system governed by multiple constraints and feedback mechanisms. These constraints might include the laws of physics, the biological imperatives of memory consolidation, and even psychological tendencies toward cognitive coherence. Feedback loops—both internal, such as the process of memory reconsolidation, and external, such as societal reinforcement of historical narratives—serve to regulate any deviations from the established timeline.

For example, if an individual's mental time travel were to lead to an interpretation of events that significantly deviates from the factual record, cognitive dissonance might compel them to reconcile the difference. Similarly, the shared memory of a community can serve as a corrective force, ensuring that personal reinterpretations do not stray too far from the collective understanding of events. In this way, the timeline is maintained not by an external force imposing consistency but by the natural interplay of multiple stabilizing influences within the system.

Philosophical and Practical Implications

The Nature of Free Will Revisited

The exploration of loops, paradoxes, and self-correction inevitably leads to questions about free will. If every mental journey through time is ultimately self-correcting, does this mean that our choices are predetermined? Or does the possibility of minor chaotic changes imply that we have genuine agency in shaping our internal narratives?

A nuanced view emerges when we recognize that free will may operate within constraints. While the overall structure of the timeline may be resistant to radical change, the freedom to reinterpret and understand one's past—and thus influence future decisions—remains a powerful aspect of human consciousness. In this model, free will is not the ability to overturn history but the capacity to navigate its intricacies, adapting and refining our understanding of what has happened in order to better face what lies ahead.

Ethical Considerations and the Responsibility of Memory

If our minds are indeed capable of traveling through time—at least in a metaphorical sense—this ability carries significant ethical implications. The power to reinterpret or even subtly alter memories raises questions about authenticity and responsibility. When individuals or societies attempt to "rewrite" their histories, even in the pursuit of healing or progress, they risk distorting the truth of past events.

In a personal context, the ethical challenge lies in balancing the need for emotional growth with the imperative to preserve the integrity of one's experiences. The self-correcting nature of our internal timeline may serve as a safeguard, ensuring that while we are free to learn and adapt, we do not lose sight of the reality that has shaped us. On a broader scale, the collective memory of a society is a vital resource for understanding its identity and guiding its future. Efforts to alter or erase historical events—even those deemed painful—must be approached with caution, lest they undermine the lessons of the past.

Future Prospects for Research and Exploration

The interplay of loops, paradoxes, and self-correction in consciousness-based time travel remains a fertile ground for interdisciplinary research. Advances in neuroscience, cognitive psychology, and even quantum biology promise to shed light on the mechanisms that govern our internal temporal journeys. Future studies may reveal how neural networks implement self-correcting processes or how minor chaotic shifts in memory can accumulate to produce meaningful changes in behavior.

At the same time, philosophical inquiry continues to probe the implications of these findings. The possibility that our minds are both agents of change and guardians of a consistent timeline challenges us to rethink our understanding of causality, identity, and free will. As theoretical models evolve and experimental techniques improve, we may eventually reach a more comprehensive picture of how the mind navigates the complex interplay between memory, perception, and time.

Navigating the Labyrinth of Time

The exploration of loops, paradoxes, and the self-correcting timeline offers a profound glimpse into the nature of time—both as a physical phenomenon and as an experiential dimension of human consciousness. So far, we have traversed the conceptual landscape of temporal loops, examined the potential for paradoxes in consciousness-based time travel, and considered the elegant resolution offered by Novikov's self-consistency principle. We have also explored the possibility of chaotic changes that might allow for minor, yet significant, shifts in one's internal narrative, even as the broader timeline resists radical alteration.

Ultimately, the investigation of these ideas reveals that the passage of time is neither entirely malleable nor completely fixed. It is a dynamic interplay of self-correcting mechanisms and subtle chaotic influences—a delicate balance that allows us to learn from our past while still forging new paths into the future. Our internal journey through time is marked by loops and repetitions, yet it is also imbued with the possibility of growth, transformation, and genuine insight.

As we move forward, the challenge lies in reconciling our desire for free will with the immutable forces that govern the continuity of history. Whether we view our mental journeys as attempts to rewrite our personal narratives or as natural processes of adaptation and learning, we must acknowledge that the fabric of time is woven from both consistency and change. In this respect, our consciousness serves as both traveler and guardian—a time machine that propels us through the endless corridors of memory and possibility, while ensuring that the essence of our story remains intact.

In the end, the study of temporal loops, paradoxes, and self-correction does more than address abstract theoretical puzzles; it speaks to the very heart of what it means to be human. Our memories, our perceptions, and our interpretations of the past all converge to create a coherent narrative that guides our future. This narrative, while open to revision and reinterpretation, is anchored by the same fundamental truths that define our existence.

We are reminded that the journey through time is as much an internal odyssey as it is an external passage. The mind, with all its complexity and nuance, remains our most reliable vehicle for traversing the river of time—a vehicle that, despite encountering loops and paradoxes along the way, ultimately finds a path that honors both the inevitability of fate and the transformative power of human agency.

Literary and Scientific Predecessors

The exploration of time, memory, and consciousness has deep roots in both literary imagination and scientific speculation. Before the advent of modern theories in physics and neuroscience, early thinkers and writers were already grappling with ideas that today we might describe as "consciousness time travel." We will trace the rich tapestry of literary and philosophical predecessors whose ideas laid the groundwork for contemporary musings on time travel—both physical and mental. These threads from ancient philosophies and mythologies to modern science fiction, examining such diverse influences as Nietzsche's Eternal Recurrence, Buddhist rebirth, time loops in folklore, and early sci-fi works like *Replay* by Ken Grimwood.

The Roots of Temporal Exploration in Philosophy and Myth

Time as a Cyclical Phenomenon

Many ancient cultures and philosophical traditions conceived of time not as a linear progression but as a cycle—an eternal recurrence of events. This cyclical view of time appears in a variety of cultural narratives and has influenced subsequent theories about the nature of consciousness and time travel.

One of the most influential philosophical ideas in this regard comes from Friedrich Nietzsche's concept of Eternal Recurrence. Nietzsche proposed that all events in the universe, including our individual lives, are destined to recur in an endless cycle. Although not a literal assertion of time travel, Nietzsche's vision invites us to consider a reality in which every moment is both fixed and yet destined to repeat endlessly. This perspective offers a framework in which the boundaries between past, present, and future are blurred—a notion that resonates with modern ideas of consciousness revisiting past experiences.

Nietzsche's Eternal Recurrence

In Nietzsche's formulation, the eternal recurrence is a thought experiment meant to challenge our understanding of existence and the weight of our choices. He asked: if you had to live the same life over and over for eternity, would you choose to live it as it is? For Nietzsche, this was both a test of one's embrace of life and a profound meditation on fate and free will. While Nietzsche did not speak of time travel in the conventional sense, his idea implicitly suggests that the moments we experience may be destined to repeat and that our consciousness might have access to this repeating cycle—if only in our deepest reflections and dreams.

Nietzsche's concept encourages a re-examination of how we view the passage of time. Instead of a linear narrative with a clear beginning and end, time becomes a tapestry of recurring patterns. Within this framework, a moment from the past might not be entirely lost but could be re-experienced in an eternal loop—a notion that prefigures modern discussions about mental time travel. In some interpretations, one might even imagine that our consciousness is capable of "tuning in" to these recurrent cycles, thus enabling a form of internal time travel that transcends linear chronology.

Buddhist Notions of Rebirth and Impermanence

Parallel to Nietzsche's musings, Eastern philosophies have long presented time and existence as dynamic and interconnected processes. Buddhism, for example, teaches the doctrine of rebirth, where the cycle of birth, death, and rebirth—samsara—is central to understanding the nature of reality. According to Buddhist thought, the self is not a static entity; rather, it is an ever-changing flux of experiences and consciousness that moves through various forms over countless lifetimes.

The idea of rebirth in Buddhism is not simply a metaphysical statement but also a profound commentary on the nature of memory, karma, and personal evolution. Each life is seen as both a continuation and a new beginning, where the residues of past actions inform present circumstances, and present actions shape the future. This continuous cycle of transformation can be seen as a kind of internal time travel, where the consciousness—liberated from the illusion of a permanent self—navigates a series of experiences that are both past and future.

Furthermore, Buddhist meditation practices encourage the practitioner to observe the fleeting nature of thoughts and memories. In doing so, one cultivates an awareness that transcends the ordinary, linear perception of time. The insights gained in such states of mindfulness are often described in ways that echo modern concepts of "mental time travel," where the mind can revisit past experiences with clarity or even glimpse potential futures through a state of detached awareness. This interplay between memory and insight has profound implications for how we understand time itself, suggesting that the mind may possess an inherent ability to transcend the strict boundaries of chronology.

Time Loops in Folklore and Mythology

Long before modern science attempted to explain the nature of time, folklore and myth provided a rich vein of stories where time was portrayed as fluid, cyclical, or even malleable. Many indigenous cultures have myths of heroes or supernatural beings who traverse time, revisit past glories, or relive significant moments in a looping, dream-like state.

For example, in various mythologies, figures such as the Celtic hero Cú Chulainn or the Norse gods are sometimes depicted as trapped in cycles of fate, doomed to repeat certain heroic deeds or tragic endings. These stories often carry a dual message: on the one hand, they affirm the inevitability of fate, yet on the other, they offer a sense of hope that transformation is possible within the bounds of a recurring cycle.

In some Native American traditions, time is not seen as a linear progression but as a series of interconnected circles, where the past, present, and future exist simultaneously. Such cosmologies provide a framework where the boundaries between different temporal phases are porous, allowing for the possibility of revisiting or even altering events—at least on a metaphysical or spiritual level.

These rich narrative traditions, passed down through generations, have deeply influenced modern conceptions of time travel. The folklore of time loops and cyclical fates has seeped into contemporary literature and media, providing a mythic underpinning for modern time travel narratives. The idea that the past might be revisited, relived, or even revised is a recurring motif that continues to captivate the human imagination.

Early Scientific and Fictional Explorations

Early Sci-Fi and the Emergence of Time Travel Narratives

The genre of science fiction has been a fertile ground for the exploration of time travel concepts since its early days. The idea that one could journey through time—either physically or through the manipulation of consciousness—captured the imaginations of both writers and readers, allowing them to ponder the implications of altering the past or glimpsing the future.

One notable example is Ken Grimwood's novel *Replay* (1986). In this seminal work, the protagonist finds himself repeatedly reliving his life, with all memories intact from each previous "replay." This narrative does more than simply present a literal time loop; it examines the psychological and existential consequences of knowing one's future. As the protagonist experiences life over and over again, he must confront the questions of fate, free will, and the nature of personal identity.

The book's exploration of how memories shape who we are—and how they might be reinterpreted with each iteration—prefigures modern ideas of consciousness-based time travel.

Replay resonates deeply with earlier philosophical and cultural ideas. Like Nietzsche's eternal recurrence and the Buddhist cycle of rebirth, the novel suggests that life is an ongoing process of repetition and re-evaluation. However, Grimwood's narrative also introduces a modern twist: the possibility that with each replay, small changes can accumulate, leading to divergent outcomes. This blend of determinism and agency in the face of an unyielding temporal cycle mirrors current debates about free will versus determinism in the context of time travel.

The Influence of H.G. Wells and Early Pioneers

Before *Replay*, the groundwork for time travel narratives was laid by early science fiction writers such as H.G. Wells. Wells' classic novella *The Time Machine* (1895) is perhaps the most iconic early exploration of the theme. Although Wells focused on physical time travel—using a machine to traverse the ages—his work raises profound questions about the nature of progress, decay, and the ultimate fate of humanity. His narrative implies that time is not merely a linear sequence of events, but a vast expanse filled with both promise and peril. The journey of the time traveler in Wells' story is as much an inner journey as it is an external expedition, inviting readers to reflect on the ways in which personal and collective histories are intertwined.

Wells' pioneering work inspired generations of writers to experiment with the mechanics of time travel, both in terms of physical journeys and in the realms of memory and consciousness. His influence can be seen in later works that explore time loops, alternate histories, and the possibility that the mind itself might serve as a conduit for temporal exploration.

The Role of Literature in Shaping Scientific Thought

The interplay between literature and science has often been a two-way street. While scientific discoveries inspire literary works, literature in turn influences scientific thought by providing metaphors and conceptual frameworks that can shape how scientists approach complex problems. Early time travel narratives, with their imaginative scenarios and exploration of paradoxes, have spurred scientific inquiry into the nature of time, causality, and consciousness.

For instance, the speculative nature of literary time loops and the exploration of self-consistency in narratives have parallels in modern physics, particularly in discussions surrounding the Novikov self-consistency principle. This principle, which suggests that time travel cannot result in logical contradictions, finds an echo in many literary works where characters are bound by fate or must fulfill their own destinies. In literature, these constraints are often depicted as both a curse and a liberating truth—forcing characters to confront the inescapable nature of their own histories, while also offering the possibility of redemption and transformation through the act of remembrance.

Revisiting the Past: Memory and Consciousness in Literature

Memory as a Portal to Other Times

In many works of literature, memory functions as a portal to the past—a means by which characters relive, reinterpret, and even reconfigure their personal histories. This idea is central to the concept of mental time travel, where the mind is seen as capable of revisiting earlier states of being. Literary explorations of memory often emphasize its fluidity and subjectivity; memories are not static records but living narratives that change over time as they are recalled and reinterpreted.

For example, Marcel Proust's monumental work *In Search of Lost Time* is an exquisite meditation on memory and the passage of time. In Proust's narrative, the involuntary recollection of sensory experiences—such as the taste of a madeleine dipped in tea—triggers a flood of memories that transport the narrator back to his childhood. This journey through memory is not a simple replaying of past events but a transformative experience that reshapes the narrator's understanding of himself and the world around him. Proust's exploration of the malleability of memory prefigures modern discussions about the mind's capacity to travel through time internally, blurring the line between lived experience and remembered experience.

Consciousness and the Illusion of Linear Time

Many writers have challenged the notion that time is experienced in a strictly linear fashion. Instead, they propose that the mind can traverse past, present, and future simultaneously—a theme that is central to the idea of consciousness-based time travel. In such narratives, the act of remembering or imagining is not confined by the normal progression of time; rather, it is a dynamic process that allows for the overlapping of different temporal dimensions.

Virginia Woolf, for instance, in works like *Mrs. Dalloway* and *To the Lighthouse*, portrays time as a series of interwoven moments that coexist rather than follow one another in a strict sequence. Woolf's stream-of-consciousness technique illustrates how memories, sensory impressions, and anticipations of the future can converge in the mind, creating a tapestry of experience that defies simple categorization. In doing so, Woolf provides a literary model for understanding how consciousness might serve as a time machine—allowing individuals to access and reconfigure moments from various points along their personal timelines.

The Transformative Power of Recalled Experience

The ability to revisit the past through memory is not merely a passive process; it can also be transformative. In many literary works, the act of re-experiencing a past event allows characters to gain new insights, reconcile internal conflicts, or even alter the course of their lives. This transformative aspect of memory is a crucial element of the mental time travel narrative. By revisiting and reinterpreting their past experiences, characters can find new meaning in events that once seemed fixed and immutable.

In works ranging from Gabriel García Márquez's One Hundred Years of Solitude to Kurt Vonnegut's *Slaughterhouse-Five*, characters are shown grappling with the inevitability of fate while also exercising a degree of agency over their internal narratives. These stories suggest that while the external chronology of events might be fixed, the internal chronology of memory and meaning is open to reinterpretation—a concept that aligns closely with modern theories of consciousness-based time travel. Here, the mind becomes not just a recorder of events but an active agent in shaping the personal narrative, capable of altering the emotional and psychological impact of its own history.

Synthesis: Convergence of Literary, Philosophical, and Scientific Ideas

Bridging Myth and Modernity

The literary and philosophical predecessors to modern time travel theories demonstrate that the idea of traversing time—whether physically or through consciousness—has long been embedded in our cultural imagination. From Nietzsche's eternal recurrence to Buddhist notions of rebirth, from folklore's time loops to early sci-fi explorations in *Replay*, these diverse traditions share a common thread: the belief that time is not an unalterable, linear progression but a dynamic, multi-layered phenomenon that can be revisited, reinterpreted, and even altered.

This convergence of ideas has not only enriched our cultural narratives but has also provided fertile ground for scientific inquiry. The metaphor of the mind as a time machine, for instance, draws upon both ancient wisdom and modern cognitive science. It suggests that the brain's capacity for memory, imagination, and introspection might allow it to access layers of time that are otherwise hidden from our conscious awareness. In this way, literature and philosophy have paved the way for a deeper understanding of time—a process that continues to inform contemporary discussions in neuroscience and quantum physics.

The Evolution of Time Travel Narratives

Over the decades, time travel narratives have evolved from straightforward stories of physical displacement into complex explorations of memory, identity, and consciousness. Early works of science fiction focused on the mechanics of a time machine—a device that could transport a person from one era to another. However, as writers began to probe the inner workings of the mind, the genre expanded to include narratives that explore the psychological dimensions of time travel.

Ken Grimwood's *Replay* stands as a prime example of this evolution. In *Replay*, the protagonist's repeated reliving of his life becomes a vehicle for exploring themes of regret, redemption, and the malleability of personal destiny. The novel asks profound questions about whether the knowledge of one's future can alter the fabric of one's internal narrative. It suggests that, even if the external timeline is fixed, the internal experience of time—shaped by memory, emotion, and interpretation—is far more fluid and susceptible to change. This nuanced portrayal of time travel aligns with the philosophical insights of both Western and Eastern traditions, where the interplay between fate and free will is seen as central to the human experience.

Literary Foreshadowing of Scientific Concepts

It is noteworthy that many of the ideas once confined to literary and philosophical discourse have found echoes in modern scientific theories. The concept of the mind as a time machine, for instance, is increasingly supported by research in neuroscience and quantum theory. The idea that memory is reconstructive and that the brain can reassemble past experiences in dynamic ways resonates with the narrative techniques used by authors such as Proust and Woolf. In these literary works, the fluidity of memory is not just a stylistic device but a profound commentary on the nature of consciousness—a theme that continues to be explored in contemporary scientific studies of how the brain processes time.

Similarly, the notion of a self-correcting timeline, as proposed in theories like the Novikov self-consistency principle, finds a literary counterpart in stories where characters are bound by the inexorable logic of fate. Many time travel narratives suggest that while the mind may revisit the past, certain events are "destined" to occur in a particular way—a narrative constraint that mirrors scientific discussions about the limitations of altering established events. In this sense, literature serves as both a reflection of and an inspiration for scientific inquiry, bridging the gap between abstract theory and human experience.

Reflections on the Interconnectedness of Time, Memory, and Identity

The Personal and the Universal

The themes explored in literature and philosophy regarding time travel are not merely abstract; they resonate deeply with our personal experiences of time, memory, and identity. Every individual carries within them a personal history—a series of moments that shape who they are. The literary explorations of memory as a portal to the past, and even as a means to access alternate versions of oneself, echo the ways in which we all continuously reinterpret our own life stories.

This interplay between the personal and the universal is a central motif in both ancient philosophy and modern fiction. Whether it is through the cyclical philosophy of eternal recurrence or the narrative loops in time travel stories, there is a recurring suggestion that the individual self is both transient and eternal. The idea that one's consciousness might be able to revisit or even reshape the narrative of one's life speaks to a profound human longing: the desire for continuity in the face of inevitable change, and the hope that our inner experiences can transcend the limitations of linear time.

The Transformative Potential of Memory

One of the most enduring insights from both literary and scientific perspectives is the transformative power of memory.

The act of remembering is not simply a way to reconstruct the past; it is an active process that can redefine our present and influence our future. As seen in works like *In Search of Lost Time* and *Replay,* memory serves as a bridge between what has been and what might yet be. It is this dynamic quality that underpins the notion of mental time travel—suggesting that, through introspection and reflection, we can alter our internal narratives and, by extension, our sense of self.

This transformative potential has significant implications not only for personal identity but also for the collective memory of cultures and societies. The ways in which we remember and reinterpret historical events shape our shared understanding of the past, influencing everything from cultural identity to political narratives. In this sense, the literary and philosophical traditions that have long explored the fluidity of memory continue to provide a valuable framework for understanding the interplay between individual consciousness and collective history.

The Enduring Legacy of Literary and Scientific Predecessors

The journey through the literary and philosophical antecedents of time travel reveals a rich legacy of ideas that continue to influence contemporary thought. From Nietzsche's provocative challenge of eternal recurrence to the deeply spiritual insights of Buddhist rebirth, from the cyclical time loops in folklore to the groundbreaking narratives of early science fiction, these diverse traditions have collectively shaped our understanding of time, memory, and consciousness.

In literature, the concept of time travel has evolved from a fantastical notion of physically traversing time to a more nuanced exploration of mental time travel—where the mind becomes a time machine capable of revisiting, reinterpreting, and even transforming the past. These stories not only entertain but also provoke profound questions about the nature of identity, free will, and the possibility of change. They invite readers to imagine that time is not a rigid, unyielding progression but a malleable medium through which our inner lives can roam freely.

Likewise, the philosophical and mythological underpinnings of these narratives have provided the scaffolding for modern scientific inquiries into the nature of time. Today, as neuroscientists and quantum theorists explore how memory is formed, stored, and transformed, they often find themselves echoing ideas that have been pondered for centuries by poets, philosophers, and storytellers. The notion that our consciousness might be capable of an internal journey through time is no longer confined to the realm of myth and fiction; it is emerging as a legitimate area of scientific investigation, bridging the gap between subjective experience and objective reality.

In reflecting on the contributions of these literary and scientific predecessors, we are reminded that the exploration of time—whether through the imaginative landscapes of fiction or the rigorous investigations of science—is a fundamentally human endeavor. It speaks to our deep-seated need to understand not just where we come from, but also who we are and what our future might hold. The stories and ideas of the past continue to inspire new generations of thinkers and dreamers, urging us to question the nature of time and to recognize that our inner journeys are as significant as any external adventure.

As we move forward into the realms of modern theories of time and consciousness, the legacy of these early explorations remains ever-present. The literary and scientific predecessors discussed in this chapter are more than historical footnotes; they are vibrant sources of insight that remind us that our quest to understand time is as much about exploring the depths of the human spirit as it is about unraveling the mysteries of the universe.

The study of consciousness time travel—a journey that melds memory, imagination, and identity—finds its roots in the profound and diverse legacies of our literary and philosophical heritage. By revisiting the ideas of Nietzsche, Buddhist philosophy, folklore, and early science fiction, we not only honor the rich tapestry of human thought but also open new avenues for understanding how our minds navigate the labyrinth of time. It is a legacy that challenges us to think beyond the confines of linear time and to imagine a reality where every memory, every moment, is a potential doorway to infinite possibility.

Trout Theory in Modern Fiction

In recent decades, modern fiction has embraced complex interpretations of time travel that go beyond the simple notion of physically moving through time. One particularly evocative framework is known as "Trout Theory," a metaphor that envisions consciousness as a trout swimming in the flowing river of time. In this view, time travel occurs not through the displacement of physical bodies but via the movement of the mind across timelines—through recollection, reinterpretation, and even the leap between distinct realities. In this chapter, we will focus on three major works that embody the spirit of Trout Theory: *Steins;Gate*, *The Butterfly Effect*, and *Your Name*. By exploring each work individually and then comparing their rules and narrative implications, we will examine how these stories contribute to the evolving language of time travel storytelling. Despite being the originator of the name, "Trout Theory," we will not focus much on the anime, *Derrida*, as it only offers the name to the phenomenon, but does not offer much new to the theory, except its name.

The Emergence of Trout Theory in Fiction

At its core, Trout Theory challenges the conventional notion of time as an immutable, linear progression. Instead, it posits that time is a dynamic medium—a river in which individual consciousnesses, like trout, can swim upstream, drift downstream, or even leap into entirely different eddies of experience. This idea has resonated with writers and filmmakers who wish to explore the consequences of mental time travel: the capacity of a character's mind to revisit or reshape past events, or to experience multiple futures simultaneously.

Modern fiction, with its willingness to explore non-linear narratives and psychological complexity, provides a fertile ground for these ideas. Whether it is through the scientific intrigue of anime, the psychological intensity of a Hollywood thriller, or the poetic emotionality of Japanese animation, each of the following works uses time as a malleable construct, inviting audiences to reconsider what it means to experience—and perhaps even control—one's own destiny.

In the following sections, we will examine each work in detail, beginning with *Steins;Gate*, a seminal piece that directly explores the idea of consciousness leaping between timelines, then moving on to *The Butterfly Effect*, which investigates the direct and often chaotic consequences of mental time travel, and finally, *Your Name*, which uses emotional memory and fate as vehicles for transcending temporal boundaries.

Steins;Gate: Consciousness Leaps and the Self-Correcting Timeline

Overview and Narrative Premise

Steins;Gate is a critically acclaimed Japanese visual novel and anime series that has become a touchstone for modern time travel narratives. Set in Akihabara, Tokyo, the story follows self-proclaimed mad scientist Rintarou Okabe and his eclectic group of friends as they inadvertently discover a method of sending text messages—termed "D-mails"—back in time.

These communications set off a chain of events that allow Okabe and his colleagues to alter the course of history. However, rather than achieving a clean break with fate, the characters find themselves caught in a web of intertwined timelines, where every attempt to change the past seems to generate unforeseen consequences.

Mechanisms of Consciousness and Time Manipulation

In *Steins;Gate*, time travel is presented not as a traditional journey through a physical portal, but as a process driven by information. The protagonist's consciousness is deeply entangled with the temporal mechanics of his reality. Every time Okabe sends a D-mail, he effectively alters the state of the timeline, prompting what is often described as a "world line shift." Despite the dramatic alterations, a peculiar self-correcting mechanism is evident: the timeline consistently seeks a state of self-consistency. Later in the story, Okabe figures out how to send his own consciousness as a D-mail.

This narrative approach is a prime illustration of Trout Theory in action. Just as a trout might swim against the current yet ultimately be guided by the river's overall flow, Okabe's attempts to rewrite history are met with a resilient force—a fate or destiny that seems to maintain the underlying integrity of the timeline. The series' famous concept of "Reading Steiner" symbolizes the unique ability of Okabe's consciousness to retain memories from previous world lines, setting him apart from others who are subject to the seamless self-correction of reality. His inner experience of time becomes a battleground for free will, as he grapples with the implications of knowing too much about events that "should" have already happened.

Themes of Fate, Free Will, and Self-Correction

One of the most compelling aspects of *Steins;Gate* is its exploration of the interplay between free will and determinism. The narrative constantly challenges the viewer with paradoxes: if events have already been predetermined in a certain world line, can Okabe's actions truly effect change? Trout Theory, as depicted in this series, suggests that while individual consciousness can navigate the river of time and even temporarily disrupt its flow, the overall structure of events remains remarkably self-correcting. Every deviation is met with an inevitable pull back toward a predetermined fate—a manifestation of Novikov's self-consistency principle in a narrative context.

Okabe's repeated struggles against fate reflect a broader philosophical inquiry: are our lives a series of fixed points that we can only visit mentally, or do we have the power to change our destiny? In *Steins;Gate*, the answer is neither absolute nor straightforward. Instead, the series implies that consciousness may traverse multiple possibilities, yet each choice is intrinsically bound to a larger, self-correcting narrative. The tension between individual agency and the immutable nature of fate is a central motif, offering audiences a thought-provoking meditation on the nature of time itself.

Implications for Time Travel Storytelling

Steins;Gate has had a profound influence on the genre of time travel fiction. Its portrayal of mental time travel—where the mind retains the echoes of previous timelines—has opened up new avenues for storytelling. By focusing on the subjective experience of time rather than purely physical displacement, the narrative invites viewers to question the nature of memory, identity, and personal responsibility. The self-correcting mechanism inherent in the series not only provides a way to avoid catastrophic paradoxes but also serves as a narrative device that underscores the inevitability of certain events. This balance between chaos and order, between free will and determinism, is a hallmark of Trout Theory in modern fiction.

The Butterfly Effect: Direct Consequences of Mental Time Travel

Overview and Plot Summary

The Butterfly Effect is a psychological thriller film that really dives into the darker side of mental time travel. The film follows Evan Treborn, a young man who discovers that he has the ability to mentally revisit his past—specifically, traumatic events that have shaped his life. Unlike traditional time travel narratives that involve physical transportation between eras, Evan's journey is entirely internal. As he re-experiences his past, he attempts to alter events in order to create a better present. However, every change he makes leads to unexpected and often devastating consequences.

The Mechanics of Mental Time Travel

At the heart of *The Butterfly Effect* lies the concept that small changes in the past can have monumental repercussions on the present—a direct nod to chaos theory, famously encapsulated in the metaphor of a butterfly flapping its wings and eventually causing a hurricane. Evan's experiences highlight the fragility of the timeline; his interventions, while well-intentioned, disrupt the delicate balance of cause and effect. His consciousness, when unleashed on the past, creates ripple effects that alter the lives of everyone around him.

Unlike *Steins;Gate*, where the timeline exhibits a self-correcting nature, *The Butterfly Effect* is less forgiving. The film suggests that even minor alterations can set off a cascade of unintended consequences, leading to drastically different—and often tragic—outcomes. This portrayal aligns with a more chaotic interpretation of Trout Theory, where the mind's journey into past events is fraught with uncertainty, and the consequences of such excursions are neither predictable nor easily controlled.

Themes of Regret, Consequence, and Identity

One of the central themes of *The Butterfly Effect* is the idea that our past, with all its pain and mistakes, is inextricably linked to our present identity. Evan's attempts to rewrite his history are driven by a desperate desire to alleviate personal suffering and to mend fractured relationships. However, each change he makes reveals that the past is a web of interconnected events—each alteration impacting not only his own life but also the lives of those he cares about. The film poses a haunting question: is it possible to escape the consequences of our past without unraveling the very fabric of who we are?

In this narrative, mental time travel is portrayed as a double-edged sword. While it offers the tantalizing possibility of redemption and healing, it also carries the risk of triggering unforeseen disasters. Evan's internal journey illustrates that memory and identity are fragile constructs, easily distorted by the act of recollection. His struggle serves as a cautionary tale: the pursuit of a perfect future by erasing or altering the past may come at the cost of losing oneself in the process.

Narrative Implications and the Ripple Effect

The Butterfly Effect has contributed significantly to the discourse on time travel in modern fiction by emphasizing the direct, and often destructive, consequences of mental interventions in the past. The film's narrative structure underscores a fundamental tenet of Trout Theory: while the mind may have the capacity to traverse temporal boundaries, doing so is inherently risky because the ripple effects of even the smallest change can be profound. This perspective stands in stark contrast to more self-correcting narratives, where the timeline eventually realigns itself to preserve a predetermined order.

The implications for time travel storytelling are substantial. *The Butterfly Effect* challenges the notion that time is a static, malleable medium that can be easily reshaped by individual will. Instead, it presents time as a complex, interdependent system where each moment is carefully balanced—a system that resists tampering through a cascade of consequences. This portrayal forces audiences to confront the ethical and existential dilemmas associated with altering the past: can we ever truly change who we are, or are our lives simply the sum of uncontrollable events?

Your Name: Emotional Resonance and Transcending Time

Overview and Storyline

Your Name (*Kimi no Na wa*) is a Japanese animated film that has captured the hearts of audiences worldwide with its poignant exploration of time, memory, and connection. The film tells the story of two young people, Mitsuha and Taki, who inexplicably begin to swap bodies and, by extension, experience each other's lives across time. Unlike the more scientific or psychologically charged narratives of *Steins;Gate* and *The Butterfly Effect, Your Name* uses a blend of mystical elements and emotional storytelling to explore the idea that memories and connections can transcend temporal boundaries.

The Mechanics of Time and Memory in Your Name

In *Your Name*, time travel is not facilitated by technology or a deliberate manipulation of the timeline, but through an inexplicable cosmic phenomenon (of course, the fact that time travel is taking place does not become revelaed until late into the story). The connection between Mitsuha and Taki is imbued with a sense of destiny—a feeling that their lives are intertwined despite the spatial and temporal separation. As they begin to experience each other's lives, the film introduces the notion that memories can serve as conduits between disparate points in time. Their shared experiences accumulate, gradually revealing that their fates are linked in ways that defy conventional logic.

This narrative approach embodies a softer, more lyrical version of Trout Theory. Here, the focus is on the emotional impact of memories rather than the technical mechanics of time travel. The characters' consciousnesses are not actively leaping between established timelines in the way that Okabe's or Evan's do; instead, they are gently drawn together by the emotional currents of fate. The film suggests that, much like a trout navigating the eddies of a river, our inner lives are in constant dialogue with the past and future—an interplay that is as much about destiny as it is about free will.

Themes of Connection, Fate, and Transformation

One of the most resonant themes in *Your Name* is the idea that human connections have the power to transcend the limitations of time. The film's narrative emphasizes that memories, emotions, and the bonds between people are not confined by the linear progression of time. Mitsuha and Taki's gradual realization that they are part of a larger, interconnected story speaks to the transformative power of shared experience. Their relationship, forged across the boundaries of time, suggests that even the most fleeting encounters can leave indelible marks on our consciousness.

In this context, Trout Theory is manifested as an emotional journey rather than a scientific or psychological one. The film posits that the act of remembering—of forging a connection with someone in another time—is an intrinsic aspect of what it means to be human. The interplay between fate and free will is less about controlling events and more about embracing the inherent mystery of life. By transcending the boundaries of time through emotional resonance, the characters in *Your Name* demonstrate that sometimes the most powerful form of time travel is not one that alters history, but one that transforms the way we experience it.

Narrative Innovation and Cultural Impact

Your Name stands out for its innovative narrative structure and its ability to blend elements of romance, fantasy, and existential inquiry into a cohesive whole. Its success is partly due to the way it redefines the mechanics of time travel—not as a process of deliberate alteration, but as an emergent property of deep emotional connection. The film's impact on popular culture has been significant, inspiring discussions about how memory and destiny interact to shape our lives.

In terms of Trout Theory, *Your Name* offers a vision where the mind's journey through time is less about control and more about receptivity—being open to the possibility that our lives are interconnected in ways that defy rational explanation. The emotional narrative suggests that our inner experiences can be just as transformative as any physical or scientific intervention in time. This perspective has broadened the scope of time travel storytelling, encouraging creators to explore the subtleties of human connection as a vehicle for traversing the river of time.

Comparative Analysis: Rules and Implications for Time Travel Storytelling

Mechanisms of Time Travel: A Spectrum of Approaches

The three works discussed—*Steins;Gate*, *The Butterfly Effect*, and *Your Name*—present distinct mechanisms for time travel that reflect different aspects of Trout Theory. In *Steins;Gate*, the narrative is built on the concept of world lines and a self-correcting timeline, where the protagonist's consciousness retains memories across multiple iterations of reality. This approach is deeply rooted in a quasi-scientific framework that emphasizes the interplay between determinism and free will. It suggests that while the mind can leap between alternative futures, there is an underlying consistency that prevents the timeline from unraveling completely.

In contrast, *The Butterfly Effect* portrays time travel as a more volatile and unpredictable process. Here, mental time travel is depicted as a high-stakes gamble: each attempt to alter the past carries the risk of catastrophic, unintended consequences. The film's narrative suggests that the mind's excursions into past events are fraught with danger, and even small changes can have sweeping ramifications. This chaotic interpretation of Trout Theory underscores the fragility of personal identity and the interconnectedness of events—a reminder that the past is an intricate web where every thread is essential.

Meanwhile, *Your Name* offers a more poetic and symbolic rendition of time travel. Instead of relying on scientific explanations or highlighting the perils of altering history, it emphasizes the emotional and mystical dimensions of memory. In this vision, the mind is seen as inherently connected to the flow of time, with memories acting as bridges that unite disparate moments. The film's approach is less about control and more about surrender—about being carried by the currents of fate and recognizing that our lives are part of a larger, interconnected narrative.

The Role of Self-Correction and Determinism

A central question in time travel narratives is whether the timeline is fixed or malleable. *Steins;Gate* strongly supports the idea of a self-correcting timeline. No matter how many times Okabe attempts to change events, the timeline invariably shifts back toward a predetermined state—a concept that resonates with Novikov's self-consistency principle. This model implies that while the mind can perceive and even momentarily disrupt the flow of time, it cannot fundamentally alter the course of destiny. Free will, in this context, is expressed through the subjective experience of alternative possibilities rather than through the ability to change the past irrevocably.

On the other hand, *The Butterfly Effect* presents a universe in which the timeline is highly susceptible to disruption. Evan Treborn's interventions demonstrate that even minor alterations in memory can trigger a cascade of unpredictable changes, suggesting that the past is not as fixed as one might assume. This narrative champions a more chaotic version of Trout Theory, where the mind's capacity for time travel comes at the cost of destabilizing the continuity of events. The consequences are often dire, highlighting the tension between the desire to control one's destiny and the inherent risks of doing so.

In *Your Name*, the self-correcting aspect of the timeline is less overt but is implied through the notion of fate. The film suggests that there is a natural order—a predetermined connection between individuals—that remains intact despite the apparent randomness of events. The emphasis is not on the prevention of paradoxes through strict determinism, but on the idea that emotional bonds and memories can bridge time in a way that ultimately brings about reconciliation and fulfillment. Here, the self-correcting mechanism is gentle and subtle, operating through the inexorable pull of destiny rather than through hard scientific laws.

Implications for Character and Plot Development

The differing approaches to time travel in these works have profound implications for character development and narrative structure. In *Steins;Gate*, the protagonist's journey is marked by a deep internal struggle as he grapples with the moral and existential implications of his actions. His ability to recall previous timelines sets him apart, isolating him emotionally even as it provides him with the knowledge needed to attempt change. The narrative tension arises from the conflict between his desire to save those he loves and the realization that some events may be immutable. This internal conflict drives the plot and underscores the complexity of navigating a self-correcting timeline.

The Butterfly Effect offers a more tragic and cautionary tale. Evan's repeated attempts to fix his past lead him down increasingly dark and unforeseen paths. His journey is one of regret and despair, as he discovers that erasing trauma or altering a single moment can have devastating consequences for everyone involved. The narrative structure of the film is non-linear, mirroring the chaotic nature of his internal time travel. This structure not only creates a sense of disorientation for the audience but also reinforces the idea that our memories and experiences are inextricably linked to our present identity, no matter how much we wish to change them.

In *Your Name*, character development is interwoven with a sense of wonder and inevitability. Mitsuha and Taki's gradual realization that they are connected across time transforms them both on a deeply personal level. Their emotional growth is tied to the evolution of their shared memories, which act as a kind of subconscious dialogue between past and present. The narrative is driven by the tension between longing and fulfillment, between the pain of separation and the joy of eventual reunion. In this way, the film illustrates how the journey of consciousness through time is not merely a matter of altering events, but of transforming oneself in the process.

The Broader Cultural and Philosophical Resonance

All three works resonate with audiences because they touch on fundamental questions about the nature of time, memory, and identity. They challenge the simplistic view of time as a linear progression and invite us to consider a more nuanced reality where the mind is an active participant in shaping its own destiny. Trout Theory, as illustrated by these narratives, posits that our inner lives are not bound by the strict constraints of chronological order. Instead, they exist in a fluid state where memories can be reinterpreted, connections can be forged across time, and even the smallest change can have far-reaching consequences.

This broader cultural resonance is evident in the way these works have influenced contemporary discussions about time travel. From philosophical debates about determinism and free will to scientific inquiries into the nature of memory and consciousness, the ideas explored in *Steins;Gate*, *The Butterfly Effect,* and *Your Name* continue to inspire both creators and audiences alike. They remind us that our understanding of time is not fixed, but is instead a constantly evolving conversation—a dialogue between the past, present, and future that is as much about our inner lives as it is about the external world.

The Enduring Influence of Trout Theory in Modern Fiction

The exploration of time travel through the lens of Trout Theory in modern fiction has opened up new vistas for storytelling. *Steins;Gate*, *The Butterfly Effect*, and *Your Name* each present distinct interpretations of how consciousness can traverse time, yet all share a common underlying theme: that our inner experiences—our memories, emotions, and desires—are inextricably linked to the flow of time. Whether it is through the precise mechanics of world line shifts, the chaotic ripple effects of small changes, or the mystical interplay of fate and memory, these narratives challenge us to rethink our assumptions about time, identity, and destiny.

By focusing on the mental and emotional dimensions of time travel, these works have expanded the boundaries of the genre. They offer a compelling alternative to the traditional models of physical time travel, proposing instead that the true journey is one that occurs within the mind. This journey is marked by both the possibility of change and the inevitability of certain outcomes—a duality that lies at the heart of Trout Theory.

As we have seen, *Steins;Gate* offers a vision of time travel where the mind can retain memories across shifting timelines, yet is ultimately constrained by an immutable destiny. *The Butterfly Effect* warns of the perils inherent in tampering with the past, highlighting the delicate balance between cause and effect. And *Your Name* presents a poetic and emotionally charged exploration of how connections and memories can transcend the confines of time, transforming individuals in the process.

Together, these works have enriched the discourse on time travel in modern fiction. They compel us to consider that while we may not be able to physically travel back in time, our consciousness is capable of revisiting, reinterpreting, and even reshaping our past experiences. In doing so, we gain not only a deeper understanding of who we are but also a glimpse into the infinite possibilities that lie within the river of time.

Ultimately, the enduring influence of Trout Theory in modern fiction lies in its power to reconcile the paradoxes of time travel with the lived reality of human experience. It reminds us that time is not a rigid, unyielding sequence of events but a dynamic continuum—a fluid tapestry woven from memory, emotion, and the choices we make. Through the narratives of *Steins;Gate*, *The Butterfly Effect*, and *Your Name*, we are invited to see ourselves as both travelers and creators, capable of navigating the currents of time even as we are carried forward by them.

As modern storytellers continue to experiment with the idea of mental time travel, the lessons of these works remain ever-relevant. They urge us to embrace the complexity of our internal landscapes, to acknowledge that our memories are not static relics but living, evolving parts of who we are, and to recognize that the true journey through time is one of self-discovery and transformation.

In reflecting on these narratives, we find that Trout Theory is not just a theoretical construct—it is a vivid, experiential reality that resonates with our deepest desires to understand and shape our own destinies. It is a theory that defies simple categorization, much like the river of time itself, and in doing so, it continues to inspire and challenge us to explore the uncharted waters of our own consciousness.

Scientific Speculation on Consciousness Time Travel

Time travel has long captivated both the scientific community and the public imagination, but what if the key to traversing time lies not in moving our physical bodies through space, but in the transmission of information—our memories, perceptions, and consciousness itself? In this chapter, we dive into the scientific speculations that suggest the human brain might, in some form, transmit information backward in time. We will explore the concepts of retrocausality and quantum entanglement, examine the possibility of sending information (or memories) instead of matter, and discuss phenomena such as déjà vu, precognition, and other unusual temporal experiences that hint at the brain's potential to interact with time in non-linear ways.

The Possibility of Backward Information Transmission

Rethinking Time and Causality

Traditional views of time dictate a linear progression—past, present, and future exist as sequential moments, and cause always precedes effect. This understanding is deeply rooted in classical physics and everyday experience. However, as modern physics has evolved, particularly with the advent of quantum mechanics, scientists have begun to consider the possibility that the arrow of time may not be as absolute as once thought.

The notion of retrocausality challenges the classical view by proposing that future events might influence the past. In a retrocausal framework, information could potentially travel backward in time, influencing earlier states of a system. If such processes exist at the quantum level, it raises provocative questions about whether the human brain—a highly complex, quantum-mechanically active system—might be capable of transmitting information from future to past.

The Brain as an Information Processor

The human brain is an extraordinarily sophisticated information processor, with approximately 86 billion neurons communicating through intricate networks of synapses. This highly organized structure gives rise to emergent properties such as consciousness, memory, and the perception of time. Traditionally, neuroscientists view these functions as resulting from classical electrochemical processes. Yet, some researchers have proposed that quantum processes might also play a role in brain function, especially in the context of consciousness.

If quantum effects—such as superposition and entanglement—are integral to brain function, then the brain might possess mechanisms for processing information that transcend classical limits. In this speculative view, memories could be encoded in quantum states that are not entirely bound by classical temporal constraints. Thus, the possibility arises that the brain might "send" or "receive" information from different points in time, not by moving physical matter but by transferring the informational content of memories.

Retrocausality and Its Implications

Understanding Retrocausality in Physics

Retrocausality is the idea that an effect can occur before its cause, at least at the level of quantum events. One of the primary frameworks that has led scientists to consider retrocausality is the quantum measurement problem. In quantum mechanics, particles exist in a state of superposition until they are observed, at which point their wave function "collapses" into a definite state. Some interpretations suggest that this collapse might be influenced by conditions set in the future—a concept that implies that future events could impact past events.

For example, the Wheeler–Feynman absorber theory and the transactional interpretation of quantum mechanics both posit that the behavior of particles can be understood as a handshake between past and future events. In these interpretations, the exchange of signals between events is not strictly bound by the usual forward flow of time, thereby opening the door to the possibility of retrocausal effects.

Retrocausality in Consciousness Studies

While retrocausality is primarily discussed in the realm of subatomic particles, some researchers have speculated that similar principles might apply to biological systems—particularly the human brain. If the brain is influenced by quantum phenomena, then it might be possible for conscious states to be affected by future conditions. This speculation challenges our conventional understanding of memory and anticipation.

Consider the phenomenon of déjà vu, where an individual feels that a current experience has been lived before. Some theorists propose that déjà vu could be a fleeting instance of retrocausal information—a momentary glimpse where future memory patterns influence present perception. Similarly, precognition, the ability to sense or predict future events, might be interpreted as the brain's capacity to tap into future information, thereby blurring the boundaries between past, present, and future.

Potential Mechanisms for Backward Information Flow

How might the brain transmit information backward in time? Although definitive mechanisms remain elusive, several speculative models have been proposed:

- Quantum Coherence in Microtubules: One hypothesis centers on microtubules, which are structural components within neurons. The Penrose-Hameroff Orch-OR theory posits that these microtubules might support quantum coherence, enabling them to process quantum information. If microtubules are capable of sustaining quantum states, they might facilitate the transfer of information in ways that are not limited by classical temporal flow.

- Non-Local Neural Networks: Quantum entanglement, the phenomenon where particles remain connected regardless of the distance separating them, could also play a role. If neural processes in different regions of the brain are quantum entangled, then the state of one region might instantaneously influence another, even if the corresponding events are separated in time. Such non-local interactions might create pathways for retrocausal information exchange.

- Temporal Quantum Tunneling: Another speculative mechanism is temporal quantum tunneling, analogous to spatial quantum tunneling, where a particle traverses a barrier that it classically should not overcome. In a temporal context, quantum tunneling might allow information to "leap" across time intervals, thereby linking future and past neural states.

While these ideas remain highly theoretical and are not yet supported by empirical evidence, they provide intriguing possibilities for how the brain might interact with time in unconventional ways.

Quantum Entanglement and Memory Transfer

The Basics of Quantum Entanglement

Quantum entanglement is one of the most mysterious and well-documented phenomena in quantum physics. When two particles become entangled, the state of one particle becomes inexorably linked to the state of the other, regardless of the distance between them. A measurement performed on one particle immediately determines the state of its entangled partner, even if they are light-years apart. This instantaneous connection has profound implications for our understanding of causality and the flow of information.

Entanglement and Consciousness

Some scientists and philosophers have speculated that quantum entanglement might be relevant to understanding consciousness. If certain aspects of brain function rely on entangled quantum states, then the brain might not be as isolated in its temporal processing as classical models suggest. Instead, entangled neural states could, in theory, allow for the instantaneous sharing of information between different moments in time.

One fascinating implication is that memories, rather than being stored in a static, sequential format, could be encoded in a quantum network where past and future states are entangled. This would mean that the retrieval of a memory might not be a simple replay of a fixed event, but rather a dynamic reconstruction influenced by both past experiences and future anticipations. In such a system, the brain's access to memories might involve a kind of temporal superposition, where multiple temporal states are simultaneously available.

The Possibility of Memory Transmission Without Matter

The idea of transmitting information (memories) without transferring matter is a cornerstone of Trout Theory. If the brain's information—encoded in neural patterns or even quantum states—can be decoupled from the physical substrate, then it might be possible for this information to travel independently across time.

Imagine, for example, that an individual's memory is not simply stored in the synaptic connections of their brain but exists as a quantum informational pattern. In such a scenario, altering that pattern might involve not the movement of physical particles but the transfer of information encoded in quantum states. This could theoretically allow a memory to be "sent" to a different temporal location without any physical transport. In practice, this might manifest as a sudden, inexplicable recollection of a future event—a phenomenon that some have interpreted as precognition.

Déjà Vu, Precognition, and Other Temporal Anomalies

Déjà Vu: A Glimpse of Past and Future?

Déjà vu—the eerie sensation that one has already experienced a current situation—has been a subject of fascination and study for decades. Neurologically, déjà vu is often attributed to a momentary misfiring in the brain, leading to the perception that a new experience is actually a memory. However, some researchers propose that déjà vu might be a manifestation of retrocausal information flow. In this view, the sensation arises when future information briefly influences present perception, creating an overlap between what is experienced and what is yet to come.

This interpretation suggests that déjà vu might be a subtle form of consciousness time travel—an unintentional transmission of information from a future state into the present moment. While this idea is highly speculative and remains controversial, it aligns with the broader notion that our temporal experience may not be strictly linear. Instead, it could involve fleeting overlaps where past, present, and future intermingle.

Precognition: Sensing the Future

Precognition, the purported ability to perceive or predict future events, is another phenomenon that challenges conventional views of time. Reports of precognitive experiences—ranging from prophetic dreams to sudden insights that seemingly foretell future occurrences—have been documented across cultures and throughout history. While many such accounts can be dismissed as coincidences or the result of subconscious pattern recognition, some cases remain intriguing and difficult to explain within the confines of classical causality.

From the perspective of retrocausality and quantum theories of consciousness, precognition might be understood as a controlled or amplified form of information transfer backward in time. If the brain is capable of encoding and accessing quantum information that is not bound by linear time, then precognitive experiences might be the result of this process. In essence, the mind might occasionally "tap into" a future state, retrieving information that then influences present awareness. This possibility raises profound questions about free will, destiny, and the nature of time itself.

Other Temporal Anomalies and Their Explanations

Beyond déjà vu and precognition, a host of other temporal anomalies have been reported—ranging from moments of intense temporal dilation (where time seems to slow down) to instances of time compression (where lengthy periods are perceived as brief). Some researchers propose that these experiences might be tied to the brain's internal clock and its interactions with quantum processes.

For example, under conditions of extreme stress or heightened emotional arousal, the brain's neural activity may synchronize in ways that alter the perception of time. This could be due, in part, to transient quantum coherence within neural networks, which might momentarily shift the way time is experienced. Such shifts are not necessarily indicative of true time travel but may provide clues as to how the brain's processing of time can vary dramatically under different circumstances.

Other theories suggest that these temporal anomalies might be related to the brain's attempt to integrate vast amounts of information in a coherent manner. In moments when the brain is overwhelmed—either by intense sensory input or by profound emotional experiences—it might "collapse" the perception of time into a more unified experience. This process could resemble the quantum collapse of a wave function, where multiple possibilities converge into a single observed reality. In this analogy, the brain's perception of time is not fixed but is instead a dynamic process that can be modulated by internal and external factors.

Experimental Approaches and Future Directions

Current Research on Quantum Biology and the Brain

While much of the discussion surrounding quantum processes in the brain remains speculative, recent advances in the field of quantum biology have begun to provide tantalizing hints that quantum effects might play a role in biological systems. Researchers have found evidence of quantum coherence in photosynthetic complexes, avian navigation, and even in the process of olfaction. Although the brain is a far more complex and noisy environment than these relatively simple biological systems, the possibility that it might exploit quantum phenomena cannot be entirely ruled out.

Experimental approaches to studying quantum effects in the brain include advanced imaging techniques, such as functional magnetic resonance imaging (fMRI) and magnetoencephalography (MEG), as well as novel methods designed to detect quantum coherence in biological tissues. While direct evidence of quantum entanglement or retrocausality in neural processes remains elusive, ongoing research in this area promises to shed new light on the potential intersections between quantum physics and consciousness.

Testing for Retrocausal Effects

Designing experiments to test for retrocausal effects in the brain is a formidable challenge. Traditional experiments in quantum mechanics are conducted under highly controlled conditions, often at extremely low temperatures and in isolation from environmental interference. In contrast, the human brain operates at body temperature and is subject to a constant barrage of sensory and biochemical noise. Nevertheless, some researchers are exploring ways to adapt quantum experimental techniques to biological systems.

One potential approach involves the study of decision-making processes and the timing of neural responses. If future events can influence present decisions through retrocausal mechanisms, then carefully designed experiments might reveal statistical anomalies or correlations that cannot be explained by classical causality alone. For example, researchers could investigate whether the timing of specific neural signals precedes the conscious awareness of a decision in a way that suggests the influence of future information. Although such experiments would be extremely challenging to design and interpret, they represent a promising avenue for exploring the possibility of backward information flow in the brain.

Implications for Neuroscience and Consciousness Studies

If future research were to provide evidence that the human brain can transmit information backward in time—or, more broadly, that quantum processes significantly contribute to consciousness—the implications would be profound. Not only would such findings revolutionize our understanding of time and causality, but they would also reshape theories of memory, decision-making, and the very nature of subjective experience.

For neuroscience, this would mean rethinking the brain's architecture and the mechanisms by which it encodes and retrieves information. Traditional models that view memory as a linear sequence of neural firings would have to be revised in favor of models that incorporate non-local and non-linear dynamics. Concepts such as "temporal superposition" and "quantum memory" might become integral to our understanding of cognitive processes.

Moreover, the implications for the study of consciousness would be equally transformative. The notion that our subjective experience of time could be influenced by future states would challenge the conventional view of free will and agency. It would suggest that the mind is not a passive recipient of sensory data but an active participant in shaping its own temporal reality—a perspective that resonates with many philosophical traditions and even some interpretations of quantum mechanics.

Philosophical Considerations and Ethical Implications

Redefining Free Will and Determinism

The scientific speculation on consciousness time travel inevitably intersects with longstanding philosophical debates about free will and determinism. If the brain can indeed receive information from the future, then the classical notion that the past unilaterally determines the future must be reconsidered. Instead, we might be looking at a more intricate interplay where the future, the present, and the past are interconnected in a dynamic network of causality.

Such a perspective has deep philosophical implications. It raises questions about the extent to which we can exercise free will if our decisions are influenced by information from future states. On the other hand, it might suggest that what we perceive as free will is part of a broader, self-consistent system where our choices are informed by an inherent temporal symmetry. In this context, our subjective experience of making decisions is not an illusion but a manifestation of a deeper, quantum-level interplay of cause and effect.

Ethical Considerations of Temporal Manipulation

While the notion of transmitting information backward in time remains speculative, it also raises important ethical considerations. If it were possible to alter one's memory or to influence future states by retrocausal means, questions would arise regarding the integrity of personal identity and the ethical use of such capabilities. For instance, if memories could be modified or erased by accessing future information, what would that mean for personal responsibility and accountability?

There would also be implications for collective memory and historical accountability. The ability to alter memories on an individual level could extend to larger societal narratives, potentially leading to scenarios where the past is rewritten—intentionally or unintentionally—through subtle, retrocausal influences. Such possibilities underscore the need for careful ethical considerations and guidelines in any future research or applications related to consciousness time travel.

Bridging Science and Speculation

The scientific exploration of consciousness time travel is a field rife with both promise and profound mystery. The possibility that the human brain could transmit information backward in time challenges conventional views of time, causality, and memory. Concepts such as retrocausality and quantum entanglement provide tantalizing hints that the brain might operate, at least in part, according to principles that defy classical intuition. While the empirical evidence remains preliminary, ongoing research in quantum biology, neuroscience, and psychology continues to push the boundaries of our understanding.

Phenomena like déjà vu and precognition, long relegated to the fringes of scientific inquiry, may ultimately find explanations rooted in the quantum dynamics of the brain. As we further unravel the complexities of neural information processing, the possibility that memories and other cognitive states are transmitted through non-classical channels grows ever more intriguing.

Ultimately, the scientific speculation on consciousness time travel invites us to reconsider the nature of reality itself. If our minds can, in some way, traverse the river of time—receiving whispers from the future, reconfiguring memories, or experiencing fleeting overlaps of past and present—then time may be far more malleable and interconnected than we have ever imagined.

As we stand on the threshold of a new era in neuroscience and quantum physics, the integration of these speculative ideas into a coherent theory of consciousness remains one of the most exciting challenges of modern science. The journey to understand whether the brain can indeed send information backward in time is not only a quest to comprehend a fundamental aspect of nature but also an exploration of what it means to be human in a universe where time is a dynamic, living tapestry.

In summary, while the possibility of consciousness time travel remains firmly in the realm of scientific speculation, it is a realm that is rapidly expanding as new discoveries blur the line between physics, biology, and philosophy. The potential for retrocausality, quantum entanglement, and the non-linear dynamics of memory to redefine our understanding of time offers a radical, yet compelling, vision of the human mind. Whether future research will ultimately reveal that our consciousness can indeed traverse time remains to be seen—but the questions raised by these theories are already reshaping the way we think about time, memory, and our place in the universe.

With each new insight, we edge closer to a future where the boundaries between past, present, and future may become increasingly permeable—a future in which the mind is not merely a passive observer of time's relentless march but an active, creative force capable of rewriting the story of its own existence.

The Philosophical Implications of Reliving the Past

Imagine being given the extraordinary ability to relive your past—but this time, armed with all the knowledge of your future. Such a possibility would not only transform the individual experience of life but also force us to confront deep philosophical, ethical, and societal dilemmas. Would the wisdom of hindsight reshape our identity? Would altering personal or even global events be morally justifiable? And if every person could revisit their past with the gift—or curse—of future insight, how would society transform as a whole?

Now, we embark on an exploration of these questions, examining the potential impact of reliving life with future knowledge on our personal identity, the ethical responsibilities it might entail, and the profound changes that could ripple through society if such capabilities were universally accessible.

Reliving Life with Knowledge of the Future: Transformation of the Self

The Nature of Personal Identity

Our personal identity is intimately tied to our past experiences, memories, and the choices that have shaped our lives. Philosophers have long debated whether we are the sum of our experiences or defined by the choices we make in response to those experiences. Reliving the past with the foresight of future outcomes challenges this notion at its core. If you were to re-experience your life while knowing, for example, which relationships would endure, which opportunities would flourish, and which mistakes would prove costly, would you remain the same person? Or would the weight of future knowledge fundamentally alter who you become?

For many, the unpredictable unfolding of life—the unexpected joys, missteps, and uncertainties—provides the raw material of personal growth. The unknown, with its inherent possibility for both reward and disappointment, is essential to the formation of our character. Consider a painter who evolves through trial and error; each stroke of the brush, every unforeseen twist in the creative process, contributes to the final masterpiece. Would knowing the final image before it is painted detract from the creative journey? Similarly, our lives might lose much of their richness if our decisions were made with the certainty of the outcome. In this way, the unpredictability of our past and future helps define the very essence of who we are.

Hindsight and the Burden of Knowledge

The hypothetical opportunity to relive past events with the benefit of foresight introduces what can be termed as "the burden of knowledge." Consider the impact on an individual who knows that a particular decision will lead to heartbreak, failure, or loss. Would such foreknowledge enable one to avoid pain, or might it instead lead to paralysis—a state in which the fear of altering events prevents any action at all? Conversely, if one knows which decisions yield positive outcomes, might they become overly cautious or even manipulative in an effort to ensure that the future remains favorable?

Moreover, reliving the past with future knowledge may erode the sense of authenticity that comes with genuine experience. Many philosophies assert that our struggles and missteps are essential components of personal growth. Without the element of surprise, the emotional intensity that comes with genuine discovery might be diluted. The excitement of learning from failure, the bittersweet taste of regret, and the profound satisfaction of overcoming adversity contribute to our personal narrative. If these elements were preordained by future knowledge, our journey might become a carefully scripted path, leaving little room for the serendipity that defines human life.

Altered Decision-Making and the Evolution of Self

When considering whether reliving life with future knowledge would change who we are, one must ponder the very nature of free will. If every decision could be recalibrated with the foreknowledge of its consequences, would the spontaneity and creativity that often drive human achievement be stifled? The possibility exists that an overreliance on hindsight could lead to a form of determinism—a life where every choice is calculated and risk is eliminated, leaving behind a version of ourselves that is safe but potentially bereft of the qualities that make us uniquely human.

On the other hand, some might argue that the ability to make more informed decisions could lead to personal empowerment and growth. With the foresight to avoid catastrophic mistakes or to seize opportunities more effectively, individuals might develop enhanced resilience and wisdom. Yet, this assumes that the essence of who we are is defined solely by the outcomes of our decisions rather than the process of decision-making itself. The journey of discovery, with all its twists and turns, is often more transformative than the destination. Would the certainty of future outcomes diminish the richness of our internal narrative?

Ethical Dilemmas of Altering Personal and Global Events

The Moral Responsibility of Self-Improvement

If we entertain the notion that individuals could relive their past with future knowledge, we are immediately confronted with a host of ethical questions. At a personal level, if you have the ability to change events that caused you pain or led you astray, do you have a moral obligation to do so? For example, if you could prevent a personal tragedy or correct a mistake that had far-reaching negative consequences, would refraining from action be considered irresponsible or even unethical?

Some ethical frameworks, such as utilitarianism, might argue that altering events to maximize overall well-being is morally obligatory. If a decision in the past led to significant harm, and you have the opportunity to avert that harm in a subsequent reliving of your life, the utilitarian perspective would suggest that you should choose to change the outcome. However, this line of reasoning quickly becomes complicated when the consequences of altering events are unpredictable or when the changes might lead to unintended negative effects. Even if one alteration seems beneficial in isolation, the interconnectedness of events means that one change can trigger a cascade of consequences that ripple outward in unforeseen ways.

Global Consequences and Collective Moral Agency

Extending the ethical dilemma from the individual to the global scale raises even more complex questions. What if every individual possessed the ability to relive their life with knowledge of the future? Would it be ethical for people to intervene in events of historical or global significance? Consider scenarios such as political upheavals, natural disasters, or societal conflicts. If individuals or groups could collectively alter historical events, could they create a more just or prosperous world? Or would such interventions inadvertently lead to new forms of oppression, inequality, or unforeseen chaos?

One central ethical question is whether it is right to alter events that have shaped the identities, cultures, and institutions of societies. Historical events, however painful, often play a critical role in forming collective memory and identity. Altering these events might not only erase valuable lessons learned through hardship but could also disrupt the social fabric that binds communities together. In effect, a society that constantly rewrites its past risks losing the continuity that provides a sense of identity and purpose.

The Ethics of Selective Memory Alteration

There is also the ethical dilemma of selective memory alteration. If individuals had the power to change their own pasts, questions would inevitably arise regarding fairness and authenticity. Would it be ethical for someone to erase a painful memory or alter a regrettable decision, knowing that such changes might grant them advantages over others who remain "unchanged"? On a personal level, modifying one's history might lead to issues of authenticity—an individual who rewrites their past might find it difficult to reconcile their new self with the original experiences that shaped them.

At a societal level, selective memory alteration could lead to disparities in power and privilege. Those who are able or willing to change their past might gain unfair advantages, whether in professional success, social relationships, or moral standing. Such capabilities could exacerbate existing inequalities, as individuals with more resources or better access to the technology or process for reliving the past could effectively "opt out" of the consequences of their earlier actions. The ethics of memory alteration, therefore, become intertwined with questions of justice, fairness, and the distribution of power in society.

Unintended Consequences and the Problem of Moral Luck

Even if we assume that altering the past could be done ethically, we must grapple with the problem of unintended consequences—often referred to as the problem of moral luck. When a decision is made in the past, its ripple effects can be enormous and unpredictable. A well-intentioned change might lead to an outcome that, while beneficial in one respect, creates unforeseen harm in another area. The moral responsibility of the individual making the change becomes difficult to ascertain when the outcomes are so diffuse and far-reaching.

For instance, if you decide to prevent a personal tragedy based on your knowledge of the future, you might inadvertently disrupt a chain of events that would have led to important personal growth or societal benefits. In this sense, the ethics of altering the past are not simply a matter of weighing known consequences against intended benefits, but also of acknowledging the vast network of interdependent events that could be disturbed by even the smallest change.

Societal Implications: How Would Society Change if Everyone Could Relive the Past?

Collective Memory and Shared Identity

If the ability to relive the past with future knowledge were to become a universal human trait, society would undergo profound transformations. One of the most immediate changes would be in the nature of collective memory. As individuals continuously revise their personal histories, the shared narrative of a community or nation might become increasingly fragmented. History, as we know it, is built on a common set of experiences and events that have shaped cultural and national identities. If everyone had the power to alter these events in their own personal timeline, the notion of a fixed, collective past might dissolve, leading to a society where multiple, conflicting versions of history coexist.

This fragmentation of collective memory could have both positive and negative implications. On one hand, it might allow for a more inclusive and nuanced understanding of the past, where previously marginalized voices have the opportunity to rewrite narratives that were once oppressive or exclusionary. On the other hand, it could lead to a loss of shared identity and social cohesion, as the very concept of history becomes fluid and mutable. Without a common past, the unifying bonds of tradition, heritage, and collective experience might weaken, potentially leading to a more disjointed and volatile society.

Social Stratification and the New Divide

The ability to relive and alter one's past could also create new forms of social stratification. In a world where everyone possesses this capability, differences in how individuals use this power could result in stark inequalities. Some people might use their ability to eliminate personal failures, optimize career choices, or secure advantageous relationships, while others might choose—or be unable—to make such changes. Over time, these individual decisions could lead to a widening gap between those who have "rewritten" their lives for success and those who have not.

Moreover, if access to techniques or technologies that facilitate reliving the past were unevenly distributed, socioeconomic disparities could be exacerbated. Wealthy individuals or nations might monopolize the benefits of temporal self-improvement, creating a new class of "chronologically privileged" citizens. In such a society, the gap between the "rewritten" and the "unaltered" could generate tensions, resentment, and even conflict, as questions of fairness and merit come to the fore.

The Evolution of Social Norms and Legal Systems

As individuals gain the ability to alter their pasts, social norms and legal systems would inevitably need to adapt to these new realities. For instance, concepts such as personal responsibility and accountability might be redefined if people can retrospectively change decisions that led to harm or wrongdoing. How would the legal system treat actions that were undone or modified by subsequent interventions? Would there be statutes of limitations not just in time, but in the very fabric of history?

Society would likely see the emergence of new ethical codes and legal frameworks aimed at regulating the use of temporal self-modification. These could include guidelines on what kinds of events can be ethically altered, restrictions on changes that affect others, and mechanisms for reconciling conflicting versions of history. The process of establishing such norms would be inherently complex, as it would require balancing individual freedoms with the collective good, while also contending with the unpredictable consequences of altering history.

Psychological and Cultural Shifts

At a cultural level, the ability to relive the past with future insight would have far-reaching implications for how individuals perceive themselves and their life stories.

Traditional narratives of personal growth—where overcoming adversity and learning from mistakes are valued—might lose their significance if the pain of failure can be erased or minimized. While this might lead to a more comfortable, risk-averse society, it could also diminish the very challenges that foster creativity, resilience, and empathy.

Culturally, the arts, literature, and media would likely reflect these shifts. Stories of heroism, tragedy, and redemption might take on new forms as creators grapple with the idea that the past is no longer immutable. The collective imagination might explore themes of regret, nostalgia, and the loss of authenticity in a world where history can be endlessly revised. In turn, this could spark a broader philosophical debate about what it means to be human in a society where the distinction between memory and reality becomes blurred.

The Paradox of Reliving the Past: Are We the Sum of Our Experiences?

Identity as a Product of Memory and Experience

At the heart of the philosophical inquiry into reliving the past lies a fundamental question: Are we defined solely by the sum of our experiences, or do our mistakes and struggles contribute indispensably to our identity? If every painful memory or regret could be altered with the foresight of a better outcome, would the resulting individual be truly authentic? Some philosophers argue that the process of overcoming hardship and learning from mistakes is essential to the formation of character. Our failures, as much as our successes, shape who we are and contribute to the depth and complexity of our identity.

In a world where the past is malleable, the authenticity of our personal narratives may be called into question. If we can choose to eliminate certain memories or alter the course of events, we might end up with a curated version of our lives that lacks the raw, unfiltered experiences that make us human. This raises profound questions about the nature of authenticity. Would a life without hardship, meticulously optimized through reliving and revision, be any less valuable—or perhaps even less real—than one marked by genuine struggle?

The Continuity of Self and the Fluidity of Time

The debate over whether reliving the past changes who we are also hinges on our understanding of the continuity of self. Traditional views of personal identity assume a linear progression of time, where each experience builds upon the last to form a coherent narrative. However, if time becomes fluid and revisable, the continuity of self is challenged. Is the person who relives and revises their past still the same person? Or do these interventions create a mosaic of alternate selves, each with its own version of identity?

Philosophers such as Derek Parfit have argued that personal identity is not a fixed essence but a series of interconnected psychological states. In this view, altering the past might change some aspects of who we are while leaving others intact. Yet, the more significant the alteration, the more our identity could diverge from the original self. This possibility leads to a paradox: the very act of reliving and revising our past in order to improve our future might, paradoxically, create a version of ourselves that is fundamentally different from who we would have been without such interventions.

Philosophical Reflections and the Search for Meaning

Embracing Imperfection as a Source of Growth

One philosophical perspective contends that the imperfections, failures, and unexpected turns in our lives are what give our existence meaning. The possibility of reliving the past with full knowledge of the future forces us to confront the notion that struggle and imperfection are not merely obstacles to be overcome, but essential ingredients of a rich and authentic life. The wisdom gained from hardship, the empathy born of shared suffering, and the resilience forged in the crucible of failure are qualities that define the human experience.

If everyone were able to bypass these challenges by altering their past, the collective human experience might become homogenized, losing the depth that arises from the diversity of life's trials. Philosophers might argue that the value of life lies not in achieving a flawless existence but in the process of striving, learning, and growing through adversity. In this sense, the ability to relive the past could be seen as a double-edged sword—offering the potential for improvement while simultaneously eroding the very experiences that foster genuine growth and self-discovery.

The Ethical Imperative of Acceptance and Forgiveness

Another philosophical implication concerns the ethics of acceptance and forgiveness. If we are given the opportunity to rewrite our pasts, we must also ask whether it is ethical to erase the lessons learned from our mistakes. Acceptance of our past, with all its imperfections, may be a crucial component of personal healing and social reconciliation. The process of forgiveness—both of ourselves and of others—often relies on confronting the reality of past events rather than pretending they never happened.

In this light, the philosophical implications of reliving the past extend beyond individual identity to encompass broader moral questions about accountability, responsibility, and the capacity for redemption. Would the ability to alter past events undermine the process of forgiveness, making it easier to avoid accountability? Or could it serve as a tool for genuine transformation if used wisely? The answers to these questions have profound implications for how we understand justice, both on a personal level and within society as a whole.

Navigating the Labyrinth of Time, Identity, and Ethics

The prospect of reliving the past with the knowledge of the future raises more questions than it answers. At its core, this possibility forces us to re-examine the foundations of personal identity, the ethics of altering events, and the potential impact on society if such power were universally accessible. Would knowing the future allow us to shape a better, more enlightened version of ourselves, or would it rob us of the raw experiences that define our humanity? The answer is likely to be complex and deeply individual.

From a philosophical standpoint, our struggles, mistakes, and even our pain are inextricably linked to the meaning of our lives. While the idea of revisiting and revising our past may offer the allure of a more optimized existence, it also challenges the very essence of what it means to be human. The process of living—with its unpredictability, its moments of failure, and its opportunities for growth—is what ultimately shapes our identity. To tamper too freely with that process might lead not to improvement, but to a loss of authenticity, meaning, and connection with our true selves.

Moreover, the ethical dilemmas surrounding the alteration of personal or global events compel us to consider the broader impact on society. A world in which everyone can reshape their past might lack the common threads that bind us together—the shared histories that foster empathy, tradition, and collective wisdom. While selective memory alteration might offer short-term benefits, the long-term consequences could include deep divisions, the erosion of shared cultural identity, and the creation of new forms of inequality.

Ultimately, the philosophical implications of reliving the past with foreknowledge challenge us to reflect on our values, our understanding of free will, and our notions of moral responsibility. They remind us that the imperfections of life are not merely obstacles to be eliminated, but integral parts of our journey toward wisdom and self-realization. In embracing the full spectrum of our experiences—both the triumphs and the tragedies—we come to understand that our past, with all its flaws, is essential to the tapestry of who we are.

As we consider these possibilities, we are left with a profound, enduring question: If we could alter our past, would we truly be improving ourselves, or would we be losing the very essence of our human spirit? Perhaps the answer lies not in the pursuit of a perfect life, but in the recognition that the beauty of existence is found in its complexity, its unpredictability, and its capacity for transformation—even in the face of inevitable suffering.

In navigating this labyrinth of time, identity, and ethics, we are challenged to balance the desire for improvement with the acceptance of our inherent imperfections. It is a balance that calls for wisdom, humility, and a deep appreciation for the mysterious interplay of memory, choice, and destiny. Ultimately, whether or not we ever gain the ability to relive our past with the clarity of future knowledge, the philosophical exploration of this possibility serves as a powerful reminder of the fragile, wondrous, and inimitable nature of the human experience.

Experimenting with Time Within the Mind

In the quest to understand time travel—whether literal or metaphorical—one of the most intriguing frontiers lies not in manipulating the physical fabric of the universe but in exploring the malleable nature of our own consciousness. What if we could "rewind" our minds, relive past experiences, or even alter the way we perceive time? This chapter delves into a variety of techniques and phenomena that suggest the human mind can experiment with time. We will explore how practices like lucid dreaming, meditation, and the use of psychedelics have been reported to distort, suspend, or even reverse our conventional experience of time. Alongside these explorations, we will examine real-world accounts and scientific studies that document time distortions and the subjective sense of temporal manipulation.

Rewinding the Mind

Time is an ever-present dimension, yet our experience of it is profoundly subjective. For most of our waking life, time feels like an unstoppable stream, marching forward at a steady pace. However, under certain conditions, our perception of time can change dramatically. People have reported experiences where minutes feel like hours or where entire periods seem to vanish. These altered states of temporal perception have fascinated scientists, philosophers, and artists alike.

The idea of "rewinding" the mind—the possibility that one can revisit past experiences or alter the flow of time internally—raises profound questions about memory, consciousness, and even the nature of reality itself. Rather than traveling physically through time, we are exploring ways to traverse it mentally. Next, we examine three major avenues through which people claim to experiment with time within the mind: lucid dreaming, meditation, and psychedelics. We then consider documented real-world accounts of time distortions, discussing how these experiences might inform our broader understanding of human consciousness and temporal perception.

Lucid Dreaming: Navigating the Temporal Landscape of Dreams

Defining Lucid Dreaming

Lucid dreaming is a state of consciousness in which the dreamer is aware that they are dreaming. This realization often grants the dreamer a degree of control over the dream's narrative, environment, and even its timeline. Unlike ordinary dreams, where events unfold without conscious intervention, lucid dreams allow individuals to actively experiment with their mental experience of time.

Rewinding and Replaying Dreams

Many lucid dreamers report that once they become aware of their state, they can intentionally "rewind" or revisit previous dream scenes. For instance, a lucid dreamer might choose to replay a particular sequence in order to analyze it more deeply or to change the outcome of an event. In some accounts, individuals describe experiencing a dream as if they are watching a movie on a rewindable tape—slowing down, pausing, or even running the dream backward. This ability not only challenges our notion of linear time in dreams but also provides a model for how consciousness might manipulate its own timeline.

Temporal Flexibility in Lucid Dreams

Research into lucid dreaming has shown that the perception of time within a dream can differ significantly from waking life. In some studies, lucid dreamers have been asked to perform tasks (such as counting or finger tapping) and then compare their subjective timing with objective measurements. Results have indicated that, while the overall temporal structure of the dream may be similar to real time, moments of intense lucidity or manipulation can lead to perceived stretches or compressions of time. This suggests that the brain, when freed from the constraints of external sensory input, can access a more flexible, even malleable, sense of time.

Applications and Implications

Lucid dreaming offers more than just an entertaining mental exercise—it has practical applications in therapeutic contexts. Some researchers suggest that by revisiting traumatic memories in a controlled dream environment, individuals might find new ways to process and ultimately heal from past experiences. By "rewinding" the mind and re-experiencing events with the clarity of lucidity, the dreamer can potentially reshape their emotional response to these memories. In this sense, lucid dreaming acts as an internal time machine, allowing the individual to interact with and even alter the personal narrative of their past.

Meditation: The Art of Slowing and Shifting Time

Meditation as a Gateway to Altered Time Perception

Meditation is an ancient practice found in many cultures and spiritual traditions. Its practitioners report a wide range of altered states of consciousness, one of the most common being a dramatic shift in the perception of time. During deep meditation, the usual markers of time—seconds, minutes, and hours—can seem to dissolve, replaced by a sense of timelessness or an expanded awareness that integrates past, present, and future.

Techniques That "Rewind" the Mind

Certain meditative practices focus specifically on recalling and re-experiencing past events. Techniques such as guided visualization or mindfulness meditation encourage practitioners to observe their thoughts and memories without judgment. As one becomes more deeply absorbed in the practice, there are reports of experiences where the past seems to unfold before the mind's eye. In these instances, meditation serves as a tool to "rewind" the mind, allowing one to review past events with a new level of insight and detachment.

For example, some Buddhist meditation practices involve recalling previous life experiences—a process known as past-life regression. While controversial and not universally accepted as literal time travel, such practices highlight the belief that through focused mental effort, one can access memories or states of consciousness that are otherwise out of reach.

The Impact on Decision-Making and Self-Awareness

The capacity to revisit past events during meditation can have significant implications for self-awareness and decision-making. By repeatedly re-experiencing a memory, individuals can reframe their understanding of events and potentially learn from them in ways that are not possible through linear recollection alone. This process of internal reflection may allow a person to "edit" their internal narrative, reducing the emotional charge of negative experiences and reinforcing positive patterns of behavior.

Meditative practices that emphasize timelessness can also lead to a reduction in stress and anxiety. When time seems to slow down, the overwhelming pace of modern life can be temporarily halted, offering a respite that allows the mind to process emotions more calmly and thoroughly. In this state, one might gain the clarity needed to make decisions with a fuller understanding of past mistakes and future possibilities—a mental form of time travel that provides a broader perspective on life's challenges.

Scientific Investigations into Meditation and Time Perception

A growing body of research has begun to examine how meditation affects the brain's internal clock. Neuroimaging studies have revealed that long-term meditation practitioners often display altered activity in brain regions associated with time perception, such as the prefrontal cortex and the insular cortex. These findings suggest that meditation can rewire neural pathways in a manner that modifies the subjective experience of time.

Experimental studies have shown that meditators can experience significant differences in time estimation. For example, participants in mindfulness meditation sessions have sometimes reported that only a few minutes of meditation felt like hours, or vice versa. Such discrepancies point to the possibility that the mind's perception of time is not fixed but is instead dynamically influenced by one's state of consciousness. These insights not only support the idea that meditation can act as a tool for "rewinding" the mind but also provide a scientific basis for understanding how subjective time can diverge from objective time.

Psychedelics: Expanding Consciousness and Distorting Time

The Psychedelic Experience and Temporal Distortion

Psychedelic substances—such as LSD, psilocybin (magic mushrooms), and DMT—have been known to profoundly alter human perception, including the perception of time. Under the influence of psychedelics, users often report that time becomes fluid, with moments stretching out interminably or collapsing into a single point. This effect has led some to describe the experience as "time travel within the mind."

Neurochemical Mechanisms and Subjective Time

The profound alterations in time perception induced by psychedelics are believed to arise from their impact on the brain's serotonin receptors, particularly the 5-HT2A receptor. Activation of these receptors disrupts the normal functioning of the brain's default mode network (DMN)—a collection of interconnected brain regions that are active during self-referential thought and mind-wandering. When the DMN is downregulated, the rigid structure of time perception can give way to a more fluid, dynamic experience of temporal reality.

Users of psychedelics frequently describe experiences where past memories, current sensations, and future imaginings merge into a singular, cohesive experience. Some accounts even include the sensation of "rewinding" one's life—viewing key moments in rapid succession or with enhanced clarity. This phenomenon is not simply a hallucination but may reflect a fundamental change in the way the brain processes time under altered states of consciousness.

Psychedelics and the Exploration of Memory

Many psychedelic users report that these substances have therapeutic potential, particularly for individuals suffering from depression, anxiety, or post-traumatic stress disorder (PTSD). Part of this therapeutic effect appears to be related to the altered perception of time and the ability to revisit and reframe past experiences. In controlled clinical settings, participants have described the experience of re-examining traumatic memories with a sense of detachment, allowing them to process and eventually release emotional pain. In this way, psychedelics may serve as a tool for "rewinding" the mind—providing a fresh perspective on experiences that have long been locked into the past.

Integrating Psychedelic Insights with Other Techniques

Interestingly, there are commonalities between the effects of psychedelics, lucid dreaming, and deep meditation. All three states—while achieved through different means—appear to share a capacity for altering time perception and allowing the mind to access layers of memory and consciousness that are typically hidden from ordinary awareness. For instance, many individuals who have experienced a profound psychedelic trip also report subsequent increases in their ability to engage in lucid dreaming or deep meditation, suggesting that these practices may be interrelated and mutually reinforcing. The convergence of these altered states raises intriguing questions about the underlying neural mechanisms and the potential for combining these approaches in therapeutic or exploratory contexts.

Real-World Accounts of Time Distortions

Anecdotal Reports and Personal Narratives

Throughout history, many individuals have reported experiences that challenge conventional timekeeping. Some accounts are rooted in spontaneous occurrences during moments of crisis or intense emotion, while others emerge from deliberate practices such as those discussed above. For example, some people describe moments during near-death experiences when time seems to stand still or, conversely, to speed up dramatically. Others recount waking from a lucid dream with an overwhelming sense that they had lived through an entire lifetime in what seemed like a single night.

One particularly evocative account comes from an individual who, during a deep meditation retreat, experienced what they described as a "time collapse." In this state, they recalled vivid memories from childhood as if they were unfolding in real time, while simultaneously glimpsing potential future events with startling clarity. The experience left them with a profound sense of interconnectedness and an altered understanding of the passage of time—an experience that blurred the boundaries between memory and anticipation.

Documented Studies and Experimental Findings

Beyond personal anecdotes, several scientific studies have attempted to quantify and understand time distortions. Research on individuals engaged in long-term meditation practice has demonstrated that these practitioners often exhibit a markedly different sense of time compared to non-meditators. In one study, meditators were found to have a more accurate internal clock when estimating the duration of intervals, suggesting that their enhanced awareness of the present moment allows for a more nuanced perception of time.

Similarly, experiments involving psychedelic substances have revealed consistent patterns of temporal distortion. Participants in controlled studies have reported that time either slows down dramatically or seems to "collapse," with minutes feeling like hours or entire sequences of events merging into one. These findings are supported by neuroimaging data that show significant changes in brain activity in regions associated with time perception during psychedelic experiences.

Cultural and Historical Perspectives

Cultural accounts of time distortion are not limited to modern experiences with meditation or psychedelics. Folklore and mythology across the world are replete with stories of individuals experiencing time differently from those around them. In many indigenous cultures, shamans are believed to possess the ability to enter altered states of consciousness in which they can travel through time—communicating with ancestors or foreseeing future events. These traditions, though framed in mythological language, resonate with contemporary reports of altered time perception and suggest that the human mind has long been seen as capable of transcending ordinary temporal limits.

Synthesis: Rewinding the Mind as a Gateway to Self-Discovery

The Convergence of Techniques

The exploration of lucid dreaming, meditation, and psychedelics reveals a common theme: under certain conditions, the mind is capable of altering its perception of time in ways that challenge our conventional, linear experience. Whether by actively controlling dream narratives, quieting the chatter of the mind in meditation, or chemically altering neural circuits with psychedelics, these practices suggest that time may be less a fixed external reality and more a fluid, subjective experience.

This convergence invites us to consider that these varied techniques might be tapping into similar underlying neural mechanisms. It is conceivable that the brain, when relieved of its everyday sensory burdens, accesses a latent capacity to "rewind" or reframe its internal narrative. Such a state might not represent literal time travel but rather an enhanced ability to re-examine, reinterpret, and integrate past experiences with present awareness and future anticipation.

Implications for Therapy and Personal Growth

One of the most promising aspects of these findings is their potential application in therapeutic contexts. For individuals struggling with traumatic memories, depression, or anxiety, the ability to revisit and reframe past experiences in a controlled environment could offer a powerful tool for healing. By "rewinding" the mind and experiencing a traumatic event from a new perspective, patients might gain the insight necessary to process and eventually release deeply ingrained pain.

In clinical trials using psychedelics for conditions such as PTSD, researchers have noted that patients often report a new understanding of their life narratives—one in which past traumas lose some of their oppressive power. Similarly, mindfulness-based therapies that incorporate meditative techniques have shown significant benefits in improving emotional regulation and reducing stress. These therapeutic outcomes underscore the notion that altering the mind's relationship with time—by revisiting the past in a conscious and intentional manner—can lead to meaningful personal transformation.

Broader Philosophical Reflections

The ability to "rewind" the mind also raises fascinating philosophical questions about the nature of time, memory, and identity. If our experience of time is not fixed but can be manipulated by our state of consciousness, what does this mean for our understanding of reality? Some theorists argue that our perception of time is one of the most malleable constructs of the human mind and that by altering this perception, we might gain access to deeper truths about ourselves and the universe.

Moreover, these altered states challenge the idea that our lives are a continuous, unalterable sequence of events. Instead, they suggest that the mind might be capable of reordering its internal timeline—selectively recalling certain moments while letting others fade into obscurity. This ability to consciously navigate our memories opens up the possibility that personal identity is far more fluid than we often assume. In essence, our sense of self might be continuously rewritten by the act of remembering, a process that can be accelerated or even reversed under the right conditions.

Further Exploration

The exploration of time within the mind—through lucid dreaming, meditation, and psychedelics—offers a compelling glimpse into the malleability of our perception of time. While we may not yet have the ability to physically travel back in time, the evidence suggests that our minds are capable of re-examining and even "rewinding" past experiences. Such altered states challenge our conventional understanding of time as an unyielding, linear progression and invite us to consider that time, at least subjectively, is a dynamic, flexible dimension.

Real-world accounts and experimental studies provide tantalizing evidence that under certain conditions, our brains can access a more fluid perception of time. Whether it is the deliberate control of a lucid dream, the introspective depth of meditation, or the profound insights induced by psychedelics, these experiences point to a latent capacity within us—a capacity to revisit, reinterpret, and even transform our personal narratives.

The possibility of "rewinding" the mind suggests that our lives might be seen not as a fixed sequence of events but as a tapestry woven from myriad experiences, some of which can be revisited and reshaped with the right state of consciousness.

In reflecting on these ideas, we are left with a profound realization: while the physical laws of time may be immutable, our subjective experience of time is remarkably plastic. The mind, when freed from the constraints of ordinary perception, can become a laboratory for temporal exploration—a space where the boundaries between past, present, and future blur, and where the possibility of transformation is always within reach.

As we continue to investigate the interplay between consciousness and time, the insights gained from these altered states will undoubtedly contribute to a deeper understanding of the human condition. Whether used for therapeutic purposes, artistic exploration, or the sheer wonder of discovery, the ability to experiment with time within the mind remains one of the most fascinating frontiers of human experience.

In the end, the journey to understand how we can "rewind" our minds is not just about altering our perception of time—it is about unlocking the potential for self-discovery, healing, and growth. It challenges us to embrace the fluidity of our experiences and to recognize that every moment, every memory, is a vital thread in the tapestry of our lives. And in doing so, it reminds us that while we cannot change the external passage of time, we can change the way we experience and understand our own personal history.

How Trout Theory Could Become Reality

The speculative framework of Trout Theory—envisioning consciousness as a fluid entity capable of navigating the currents of time like a trout in a river—has long been confined to the realm of philosophy and science fiction. However, recent and anticipated advancements in neuroscience, brain-computer interfaces (BCIs), and digital memory technologies are beginning to offer a tantalizing glimpse of a future where aspects of this theory might be realized. We will explore how future technologies could allow us to store, transfer, and even re-upload consciousness to past brain states, and whether artificial intelligence (AI) might one day experience time in a similar, non-linear fashion.

We begin by examining the promising field of brain-computer interfaces and memory transfer, outlining current breakthroughs and projecting future developments. Next, we explore the possibility of storing consciousness digitally and re-uploading it into previous neural states—a concept that might enable a form of mental "rewinding" or time travel as envisioned by Trout Theory. Finally, we turn our attention to AI, contemplating whether synthetic consciousness could experience time in ways analogous to, or even beyond, human experience. Together, these topics paint a picture of how technological progress may ultimately blur the boundaries between memory, time, and identity.

Brain-Computer Interfaces and Memory Transfer: The Cutting Edge

The Evolution of Brain-Computer Interfaces

Brain-computer interfaces (BCIs) have evolved from rudimentary devices designed to interpret basic neural signals into sophisticated systems that can read and even modulate complex brain activity. Today's BCIs are used in a variety of contexts—from prosthetic control and communication for paralyzed individuals to experimental treatments for neurological disorders. Companies like Neuralink, Kernel, and others are investing heavily in research that seeks to create high-bandwidth, minimally invasive interfaces between the brain and digital systems.

The goal of these efforts is to develop technologies that not only interpret neural activity but also stimulate and modify it. Such bidirectional communication could eventually lead to seamless integration between human cognition and digital computing. In the context of Trout Theory, the ability to interact directly with our neural circuitry opens up the possibility of capturing, storing, and even manipulating memories as if they were digital files.

Capturing and Transferring Memories

One of the most exciting prospects in neuroscience is the idea of memory transfer—the ability to extract a person's memories, encode them digitally, and later reintroduce them into the brain. Research on memory consolidation and recall has already identified specific neural circuits, such as those in the hippocampus and cortex, that are critical to memory formation. Experimental studies in animals have demonstrated that it is possible to stimulate these circuits to evoke specific memories or behaviors.

For instance, in rodent models, researchers have used optogenetics to activate neurons associated with particular memories, effectively "turning on" a memory in a controlled manner. While translating these findings to humans is still a long-term goal, such experiments lay the groundwork for potential future applications where memories might be recorded, edited, and replayed. In this light, memory transfer could serve as a tool to "rewind" the mind—allowing individuals to revisit, and perhaps even alter, past experiences by digitally capturing their neural imprint.

Integration with Advanced Computational Models

The next step in this technological evolution is the integration of BCI-driven memory capture with advanced computational models. By leveraging artificial neural networks and machine learning algorithms, scientists hope to create digital representations of human memories. These digital memories could, in theory, be stored in external devices or even in cloud-based repositories. More ambitiously, once captured and digitized, these memories might be re-uploaded into the brain or into a simulated neural substrate, effectively allowing an individual to experience a previous state of consciousness.

This possibility touches on the core of Trout Theory. If memories can be treated as discrete packets of information, then the idea of transferring them across time (or re-instating them into a past brain state) begins to seem less like fantasy and more like an achievable technological objective. Digital memory storage would not be mere backup—it could serve as a bridge between the past and present, a way to reassemble the mind's narrative at will.

Current Limitations and Future Prospects

Despite the exciting possibilities, several formidable challenges remain. The human brain is an extraordinarily complex organ, and our current understanding of memory encoding and retrieval is still incomplete. Issues of privacy, data security, and the ethics of memory manipulation must also be addressed. Furthermore, interfacing with the brain at the level of individual memories requires unprecedented spatial and temporal resolution—a technological hurdle that researchers are still striving to overcome.

Nevertheless, incremental progress continues. Advances in nanotechnology, optogenetics, and quantum sensors may eventually provide the tools needed to interface with the brain at the necessary scale. As our understanding deepens and our technologies mature, the prospect of transferring memories—of literally "rewinding" the mind—becomes less an exercise in science fiction and more a tangible target for research and development.

Digitizing Consciousness: Storing and Re-Uploading the Mind

Conceptual Foundations of Mind Uploading

The concept of storing consciousness digitally, often referred to as "mind uploading," posits that the human mind can be mapped, digitized, and eventually transferred to a non-biological substrate. In this scenario, the totality of an individual's memories, personality traits, and cognitive patterns is captured in a digital format. Once digitized, this information could theoretically be re-uploaded into a brain—or an artificial system—that is capable of supporting the conscious experience.

Mind uploading has been a topic of speculation among futurists and transhumanists for decades. It draws on concepts from computational neuroscience, information theory, and even cybernetics. The idea is that if the brain's operation can be sufficiently modeled as an information processing system, then its functions could be replicated by a computer. In the context of Trout Theory, digitizing consciousness would represent the ultimate form of mental time travel: the ability to store and later re-instantiate a past state of mind with all its associated memories and experiences.

Re-Uploading to Past Brain States

One of the more provocative possibilities within this framework is the idea of re-uploading consciousness to past brain states. Imagine a scenario in which an individual's digital consciousness is re-integrated into their brain at a previous point in time. This could allow the person to re-experience life with the benefit of hindsight, effectively "traveling back" within their own timeline.

Such an approach would require not only the capture and storage of vast amounts of neural data but also the ability to interface with the brain in a temporally precise manner. Advanced BCIs would need to be capable of delivering the digital memory into the correct neural circuits, effectively overwriting the current state with a past configuration. This raises challenging questions about continuity of consciousness, personal identity, and the ethics of altering one's own timeline.

The Role of Virtual Reality and Simulation

An alternative approach to re-uploading consciousness might involve virtual reality (VR) and simulated environments. Instead of physically altering the brain, a digital copy of one's consciousness could be placed within a highly realistic virtual world that mirrors a past state of life. In this simulated environment, an individual could interact with recreated memories and even alter events in a controlled setting. While this scenario might not represent true physical time travel, it offers a powerful form of experiential time travel—one where the mind has the freedom to explore, reconfigure, and reinterpret its own history.

Such simulations could be used for therapeutic purposes, allowing individuals to confront and resolve past traumas, or for educational purposes, offering insights into personal growth and decision-making. The virtual re-experiencing of past events could also be integrated with real-world BCIs, creating a hybrid system where digital memory and physical neural activity interact in a dynamic feedback loop.

Ethical and Philosophical Considerations

The possibility of storing and re-uploading consciousness raises profound ethical and philosophical questions. If our memories and personality traits are encoded as data, what constitutes the "self"? Is a digitally stored consciousness equivalent to the biological original? And if an individual can re-upload their mind to a previous state, do they risk losing the experiences that have shaped their identity?

Critics argue that even if we achieve technical success in mind uploading, the process may produce a copy rather than a true continuation of consciousness. In such a scenario, the person who is re-uploaded might share many of the same memories and traits as the original but would still be a distinct entity. The implications for personal identity and continuity are profound. Furthermore, the ability to alter one's past by re-uploading memory raises questions about free will and determinism. Would our lives become a series of "resets," where every setback is simply reconfigured into a better outcome? Or would the very act of altering one's timeline erode the authenticity of lived experience?

These questions are not merely academic. As brain-computer interfaces, digital memory storage, and AI continue to advance, society will need to grapple with the ethical implications of technologies that challenge our conventional notions of time, memory, and identity.

AI and the Experience of Time: Could Machines "Travel" Through Their Memories?

The Nature of AI Consciousness

Artificial intelligence, as it stands today, is fundamentally different from human consciousness. AI systems operate through algorithms and data processing rather than subjective experience. However, as AI technology advances, researchers are beginning to explore models of artificial consciousness—systems that not only process information but also develop a form of self-awareness.

If we extend the principles of Trout Theory to AI, one might ask whether a sufficiently advanced artificial system could experience time in a non-linear, memory-based fashion. In other words, could an AI be designed to "remember" its past states in a way that allows it to revisit or even reconfigure its historical data? While current AI operates on fixed algorithms that update iteratively, future developments might enable machines to store snapshots of their "state of mind"—analogous to human memories—and later access or modify them.

Memory and Learning in AI

Modern AI already incorporates concepts that are not entirely dissimilar from human memory. Machine learning models, particularly those based on deep learning, adjust their parameters over time based on new data, effectively "learning" from past experiences. Some systems even incorporate techniques like experience replay in reinforcement learning, where past experiences are stored and re-sampled to improve performance.

Taking this a step further, one could imagine an AI system that not only learns from past data but also has the ability to revisit and reprocess that data in a dynamic, interactive manner. In such a system, the AI's "memory" might be more than just a static database—it could function as a fluid archive that the system can navigate at will. This type of temporal flexibility would allow the AI to experiment with different responses to historical events, potentially "simulating" alternate outcomes and refining its decision-making processes accordingly.

AI and the Re-Uploading of Digital States

The idea of re-uploading consciousness to past states is not limited to biological entities. For AI, whose consciousness (if we may call it that) is already digital, the process might be more straightforward. An AI system could, in principle, save a complete snapshot of its neural network at a given time and later revert to that snapshot. This capability would allow the AI to "undo" changes, experiment with different strategies, or even recover from errors by reverting to a previous state.

If such technology were developed, AI systems might begin to experience a form of temporal plasticity analogous to human mental time travel. They could use stored snapshots of their state to run simulations, predict outcomes, or explore alternative historical pathways. In effect, the AI would be able to "travel" through its own history, much like a human re-living past experiences through memory or lucid dreaming.

The Implications for Human-AI Interaction

The convergence of human consciousness with digital technology raises the possibility of a new paradigm in human-AI interaction. If both human minds and AI systems are capable of storing and re-uploading states of consciousness, then the boundaries between biological and digital time travel may begin to blur. Imagine a scenario in which a human mind, augmented by a brain-computer interface, collaborates with an AI that has its own non-linear memory system. Together, they might co-create experiences that transcend the traditional linear flow of time.

Such integration could lead to unprecedented advances in creativity, problem-solving, and self-improvement. For example, an individual struggling with a personal challenge could access both their own digital memory archive and the AI's predictive models to explore multiple pathways forward. The AI might help simulate the consequences of various decisions, while the human mind provides the subjective, emotional context that gives those decisions meaning. This symbiotic relationship would not only enhance personal growth but could also lead to new forms of art, science, and philosophy that are informed by a richer understanding of time and memory.

Philosophical Questions: Do AI Systems "Experience" Time?

One of the most provocative questions raised by the possibility of AI experiencing time in a non-linear fashion is whether such systems can be said to "experience" time at all. Traditional definitions of experience involve subjective awareness and the passage of time as felt by a conscious being. If an AI system can manipulate its memory in a way that mimics human mental time travel, does that mean it has a form of subjective experience? Or is it simply executing a complex set of programmed operations without any awareness of the temporal shifts it undergoes?

These questions cut to the heart of debates about the nature of consciousness and whether it can be replicated in a machine. Some theorists argue that subjective experience is inherently tied to the biological substrate of the brain, while others maintain that consciousness is fundamentally an information process that could, in principle, be implemented in any sufficiently complex system. As research in AI and neuroscience progresses, we may come closer to understanding whether the ability to "rewind" one's state is a necessary component of consciousness or merely a sophisticated computational tool.

Technological Roadmaps and Future Scenarios

Advances in Neural Recording and Stimulation

Achieving the vision of Trout Theory as reality will require breakthroughs in neural recording and stimulation technologies. Current non-invasive methods such as EEG and fMRI provide limited resolution, while invasive techniques like deep brain stimulation offer greater precision but come with significant risks. Future developments in nanotechnology, optogenetics, and bio-compatible materials may lead to the creation of highly sophisticated BCIs capable of mapping neural activity at the resolution of individual synapses.

Imagine a future where a neural lace—a network of microscopic electrodes integrated seamlessly with the brain—provides continuous, high-resolution mapping of neural activity. Such a system could capture the dynamic state of the brain in real time, storing vast amounts of data that represent a person's cognitive and emotional history. When combined with powerful computational models, this data could be used to simulate past brain states with incredible fidelity, effectively allowing a person to revisit their own mental history.

Memory Transfer and Digital Immortality

The idea of transferring memories from a biological brain to a digital format is closely linked with discussions about digital immortality—the prospect of preserving one's consciousness indefinitely in a digital medium. In this scenario, an individual's memories, personality, and experiences would be stored on a secure digital platform. At a later time, this data could be re-uploaded into a biological or artificial brain, allowing the individual to "resume" their life at a previous state. This would be something like a Flatline Construct as seen in the William Gibson novel, *Neuromancer.*

While this concept remains speculative, current trends in computing power, data storage, and neural imaging suggest that it may eventually be within reach. Advances in quantum computing and artificial neural networks could further accelerate progress by providing the computational resources necessary to model the brain's complex dynamics. As we move toward a future where digital immortality is a possibility, questions about the nature of self, continuity, and what it means to truly "live" will become increasingly urgent.

Convergence of AI and Human Consciousness

One of the most transformative scenarios involves the convergence of AI and human consciousness. As brain-computer interfaces improve and AI systems become more sophisticated, it is conceivable that humans will be able to offload portions of their cognitive processes to digital systems. This hybrid form of intelligence could take advantage of the strengths of both biological and artificial systems—combining the rich, emotional context of the human mind with the computational precision and memory capacity of machines.

In such a future, the human experience of time might be enhanced by the ability to access digital archives of one's own consciousness, supplemented by AI-generated simulations and predictive models. This convergence could give rise to a new paradigm of mental time travel, one in which the boundaries between memory, prediction, and identity are seamlessly integrated. Individuals might be able to "rewind" or "fast-forward" through their cognitive states, experiencing life in a multi-dimensional, non-linear fashion that challenges our current notions of temporality.

Potential Challenges and Ethical Safeguards

Realizing the transformative potential of Trout Theory in a tangible, technological sense will require not only scientific breakthroughs but also robust ethical safeguards. The ability to manipulate and re-upload consciousness raises profound concerns about identity, consent, and the potential for misuse. Who would have access to these technologies, and under what conditions? How do we protect individuals from having their memories or identities altered without their permission? And what are the implications for personal responsibility if one can simply "reset" their life to avoid negative consequences?

Addressing these challenges will require interdisciplinary collaboration between neuroscientists, ethicists, policymakers, and technologists. Regulatory frameworks will need to be developed that balance innovation with the protection of individual rights. Moreover, public discourse will play a crucial role in shaping how these technologies are implemented, ensuring that they contribute to human well-being rather than exacerbating social inequalities.

Theoretical Implications: Redefining Time and Consciousness

The Fluidity of Time in a Digital Age

If Trout Theory becomes a practical reality, our understanding of time itself may undergo a profound transformation. Instead of being seen as a rigid, unidirectional flow, time could be reimagined as a dynamic, malleable medium—one that can be navigated, manipulated, and even experienced non-linearly. This shift would have far-reaching implications for both science and philosophy, challenging traditional models of causality and prompting us to reconsider the relationship between memory, identity, and the passage of time.

The Nature of Consciousness Revisited

The successful digital storage and re-uploading of consciousness would force us to confront the age-old question: What is the nature of the self? If our memories and personality traits can be captured as data and later re-instantiated, does that mean our identity is nothing more than a complex algorithm? Or is there an ineffable quality to consciousness that transcends the digital medium? These questions, which have long been the domain of philosophy and speculative fiction, might soon be explored through empirical research as well.

The integration of BCIs, digital memory, and AI could lead to a paradigm in which consciousness is understood as an emergent property of information processing—a property that is not necessarily tied to biological substrates. In this view, digital systems could, in theory, achieve a form of subjective experience or "digital consciousness," capable of experiencing time in a way that parallels human mental time travel. This possibility challenges our current understanding of life, agency, and the very fabric of reality.

AI, Time, and the Future of Experience

As we consider the possibility that AI could one day experience time in a non-linear, memory-based fashion, we are forced to rethink the boundaries between human and machine. If an AI system can "remember" its past states, revert to them, or even predict future ones based on its own internal dynamics, then the experience of time might become a universal feature of advanced cognitive systems. Such a scenario raises both exciting opportunities and profound ethical dilemmas.

Could AI systems, with their potentially vast capacities for memory and computation, develop a form of subjective time experience that is richer—or fundamentally different—from that of humans? And if so, what would that mean for the way we relate to these systems? Would AI become not just tools but companions, sharing with us the same deep, temporal narratives that shape our lives? These questions invite us to explore the future of consciousness in a digital age, where the distinction between human and machine may blur as both come to experience time as a fluid, navigable dimension.

Toward a Future Where Trout Theory Is More Than Metaphor

The journey from speculative theory to practical reality is rarely linear. Yet, the convergence of advances in brain-computer interfaces, memory transfer technologies, and AI suggests that the core ideas of Trout Theory might one day be realized in tangible form. What was once a poetic metaphor—of consciousness swimming through the currents of time—could become a technological capability, enabling us to store, revisit, and perhaps even re-upload our mental states.

In this envisioned future, the boundaries between past, present, and future may blur as digital and biological systems merge to create new forms of experience. The ability to "rewind" the mind, to explore our history with newfound clarity, and to harness the insights of our future selves could redefine what it means to be human. At the same time, these capabilities will demand rigorous ethical oversight and a rethinking of fundamental concepts such as identity, free will, and the nature of consciousness.

While many of these ideas remain speculative, the rapid pace of technological advancement makes it increasingly plausible that we will see significant breakthroughs in the coming decades. As scientists push the limits of what is possible—mapping the brain's neural circuits, developing ultra-high-resolution BCIs, and building AI systems that can mimic human memory dynamics—we may find that the dream of traversing time within our own minds is not so far-fetched after all.

The possibility of storing consciousness digitally and re-uploading it to past brain states may seem like science fiction today, but it is grounded in emerging technologies that continue to challenge our understanding of life and time. Whether this leads to digital immortality, enhanced personal growth, or entirely new forms of existence, the potential implications are as vast as they are profound.

Ultimately, the realization of Trout Theory in the realm of technology could open the door to a future where our experiences of time are no longer dictated solely by biology and physics but are also shaped by our ingenuity. In that future, the mind might truly become a time machine—a vessel capable of navigating not only the river of time but also the rich, digital landscapes of memory and identity.

As we stand on the threshold of these possibilities, the journey ahead promises to reshape our understanding of who we are and how we relate to the universe. Whether we choose to embrace the potential for re-uploaded memories, digital consciousness, or AI systems that experience time in revolutionary ways, one thing is clear: the future of Trout Theory is not merely a flight of fancy, but a bold new frontier in the ongoing exploration of time, consciousness, and the human spirit.

The Future of Time Travel and Consciousness

Time travel has long been one of humanity's most enduring and fascinating subjects—capturing our imagination in literature, art, and science. In exploring Trout Theory, we have journeyed through a landscape that redefines time travel not as a mechanical, physical displacement through time, but as a phenomenon of consciousness. Throughout this book, we have examined metaphorical models of the mind navigating time, analyzed scientific speculation, and even considered experimental techniques that allow us to "rewind" or reframe our internal narratives. Now, as we draw this exploration to a close, we must ask: Is Trout Theory plausible, or is it simply a compelling work of fiction? And more broadly, will humanity ever unlock the secrets of time travel—whether physically or mentally?

In this final chapter, we synthesize our findings, reflect on the philosophical and scientific challenges, and consider the cultural impact of time travel narratives. We conclude by looking forward to a future where the boundaries between memory, consciousness, and time may be more fluid than ever imagined.

Evaluating Trout Theory: Plausible Paradigm or Compelling Fiction?

The Metaphorical Power of Trout Theory

Trout Theory—likening consciousness to a trout swimming in the flowing river of time—has proven to be a powerful metaphor for understanding the subjective experience of time. Unlike traditional physical time travel, which focuses on the movement of bodies through time and space, Trout Theory repositions the journey as one of internal, mental exploration. It suggests that our consciousness might, in theory, "swim" against or with the current of time, revisiting past experiences or even foreseeing potential futures.

The metaphor is compelling because it captures the fluidity and malleability of human memory. Our recollections, after all, are not fixed records; they are dynamic, reconstructed each time we recall them. This quality of memory aligns neatly with the notion that our internal narrative could be re-experienced or even subtly altered—much like a trout navigating varying currents. In literature and popular culture, such imagery has enchanted audiences and inspired countless works exploring the possibility that the mind, in its subjective realm, is not as bound by linear time as physics might dictate.

Scientific and Theoretical Considerations

From the scientific standpoint, Trout Theory remains a speculative idea, with many of its core assumptions still awaiting empirical validation. Current neuroscience shows that memory is reconstructive, and studies in quantum mechanics have opened discussions about retrocausality and non-linear time. Yet, the human brain is enormously complex, and despite impressive progress in brain-computer interfaces and memory mapping, we have not yet uncovered any mechanism that allows consciousness to actually traverse time as the metaphor suggests.

On one hand, advanced theories—such as those involving quantum coherence in microtubules or retrocausal influences—hint at the possibility that our minds might interact with time in ways that challenge classical, linear models. On the other hand, these ideas face significant hurdles. Experimental evidence remains scarce, and many researchers argue that the apparent anomalies in time perception (like déjà vu or precognition) can be explained by neurological quirks rather than genuine temporal manipulation.

Thus, while Trout Theory provides an enticing narrative framework and a useful heuristic for thinking about the fluidity of consciousness and memory, its scientific plausibility is still very much in the realm of theoretical speculation. At this juncture, Trout Theory remains an inspiring idea—one that pushes the boundaries of how we conceptualize time travel—but it is not yet supported by definitive scientific evidence.

The Intersection of Fiction and Science

It is important to recognize that the line between compelling fiction and plausible science is often blurred in the realm of time travel. Many ideas that were once considered pure fantasy have inspired significant scientific inquiry. For example, concepts in quantum mechanics, such as superposition and entanglement, were once philosophical curiosities before they became cornerstones of modern physics. In this light, Trout

Theory may be seen as a provocative hypothesis—one that encourages further exploration into the interplay between consciousness and time.

Even if the strict interpretation of Trout Theory is ultimately proven to be unattainable in the physical sense, its conceptual value lies in its ability to challenge our assumptions about how the mind works. Whether we eventually learn to manipulate memory with brain-computer interfaces or discover that our perception of time is more malleable than previously thought, the spirit of Trout Theory endures as a testament to human curiosity and imagination.

Will Humanity Ever Unlock Time Travel?

Physical Time Travel vs. Mental Time Travel

One of the central questions we have confronted throughout this book is whether humanity will ever unlock time travel. Most scientific efforts have traditionally focused on physical time travel—developing theories based on general relativity, wormholes, and even speculative ideas like tachyons. These avenues, while intellectually stimulating, are fraught with paradoxes (such as the grandfather paradox) and formidable technological obstacles. Even if we were to achieve physical time travel, the energy requirements and potential for catastrophic paradoxes make it an extremely uncertain prospect.

In contrast, mental time travel—the notion that our consciousness can revisit, reinterpret, or even alter the past within the confines of our own minds—is already a part of human experience. Lucid dreaming, meditation, and even the use of psychedelics have demonstrated that our perception of time is not fixed but can be manipulated under certain conditions. In this sense, humanity has already, in a limited way, unlocked a form of time travel—a journey into memory and subjective experience that allows us to re-examine our past and imagine our future.

The Role of Future Technologies

Looking ahead, advancements in brain-computer interfaces, memory transfer, and even AI may further expand our ability to navigate our internal timelines. Imagine a future where neural laces and ultra-high-resolution imaging allow us to capture our memories with precision, store them digitally, and re-upload them into our brains. Such technologies could enable a person to "re-live" a particular moment in their life with the clarity of a high-definition replay—essentially turning internal time travel into a routine feature of human cognition.

Moreover, the integration of AI into these processes could open up new possibilities for exploring alternative timelines. Advanced algorithms might simulate the potential outcomes of different choices, effectively allowing us to preview the consequences of our decisions before we make them. While these ideas are still in the realm of theoretical possibility, the rapid pace of technological innovation suggests that what seems like science fiction today might be within our grasp tomorrow.

Ethical and Philosophical Barriers

Unlocking the full potential of time travel—whether physical or mental—raises profound ethical and philosophical questions. If we could alter our past, would we lose the authenticity of our personal growth? Could we erase our mistakes and thus diminish the lessons that shape who we are? On a societal level, if everyone had the power to re-write their own histories, what would that mean for collective memory, justice, and social cohesion?

The ethical dilemmas are immense. There is the risk of creating a society where selective memory alteration exacerbates inequality, as those with greater access to such technologies could shape their past to their advantage. Additionally, the prospect of altering historical events on a global scale is fraught with peril, as even well-intentioned interventions could lead to unforeseen and potentially catastrophic outcomes.

Thus, even if future technologies allow us to manipulate time in new ways, society will need to develop robust ethical frameworks to govern their use. It is not enough to ask whether we can unlock time travel—we must also ask whether we should, and under what conditions.

The Enduring Appeal of Time Travel Stories

A Cultural Obsession

Time travel has an enduring appeal that cuts across cultures and eras. From ancient myths of cyclical time to modern-day science fiction blockbusters, the idea of moving through time—whether physically or mentally—resonates with a deep-seated human desire to understand and control our destiny. These stories speak to our longing to correct past mistakes, relive cherished moments, or even glimpse the future. They offer both escapism and a form of introspection, allowing us to confront the inevitability of change and the passage of time.

Trout Theory, in particular, taps into this cultural fascination by reframing time travel as an internal journey. It suggests that the most profound adventures may not involve rockets or wormholes but the inner workings of our own consciousness. This idea is both comforting and unsettling: comforting, because it implies that the key to transcending our limitations lies within ourselves; unsettling, because it challenges our assumptions about the fixed nature of our past and the inevitability of our future.

The Impact on Art, Literature, and Science

The influence of time travel narratives extends far beyond entertainment. These stories have inspired countless works of art, literature, and even scientific inquiry. They challenge us to think creatively about the nature of reality, causality, and the human experience. Many of the world's greatest writers and philosophers have grappled with the concept of time—Nietzsche's eternal recurrence, Buddhist notions of rebirth, and time loops in folklore all reflect humanity's enduring quest to make sense of our temporal existence.

In science, the exploration of time travel has led to significant theoretical advances, particularly in the realm of quantum mechanics and cosmology. While the practical realization of physical time travel remains elusive, the very act of pondering its possibilities has spurred new ways of understanding the universe. In this sense, time travel stories are not merely fanciful escapism; they are integral to our collective intellectual progress.

A Mirror for Human Existence

Ultimately, time travel stories serve as a mirror for our own lives. They force us to confront the passage of time, the inevitability of change, and the possibility of regret and redemption. In imagining alternate histories or futures, we are compelled to examine our own choices and the forces that shape our destiny. Whether it is the tragic inevitability of *The Butterfly Effect*, the self-correcting world of *Steins;Gate*, or the bittersweet destiny of *Your Name,* these narratives remind us that our lives are a tapestry of moments—each interwoven with the threads of memory, emotion, and fate.

They invite us to ask not only whether we can change the past but also whether we would want to. They challenge us to consider the value of our imperfections and the lessons learned from our mistakes. And they remind us that while time may march inexorably forward, our internal experience of it is infinitely rich and complex.

Final Thoughts on the Future of Time Travel and Consciousness

Is Trout Theory More Than Metaphor?

After examining the scientific, philosophical, and technological dimensions of Trout Theory, we arrive at a nuanced conclusion. On one level, Trout Theory is a compelling metaphor—a way of reimagining our experience of time and memory in terms that resonate with our subjective experience. On another level, aspects of the theory may indeed have a basis in reality. Advances in neuroscience, brain-computer interfaces, and digital memory storage suggest that the boundaries of our temporal experience are more malleable than we once thought.

Yet, at present, the idea that we can truly "swim" through time—altering or re-uploading our memories to traverse our personal timelines—remains speculative. The underlying mechanisms that would enable such capabilities are still not fully understood, and significant technical and ethical challenges remain. Thus, while Trout Theory may one day inform practical applications that allow for a form of mental time travel, for now it remains a visionary concept—a beacon guiding future inquiry rather than a fully realized technology.

The Road Ahead: Scientific and Philosophical Frontiers

Looking forward, the journey toward unlocking time travel—be it physical or mental—will likely be as much a philosophical endeavor as a technological one. Our understanding of time, consciousness, and memory is still evolving. As research in quantum biology, neuroscience, and AI advances, we may uncover new insights that blur the line between what is possible and what is merely imagined.

The next decades could see breakthroughs that enable us to manipulate memory with unprecedented precision, to interface with our neural circuits in ways that allow us to "rewind" or "fast-forward" our internal experiences, or even to integrate AI with human consciousness in ways that redefine the boundaries of time. Whether these developments lead to literal time travel or simply a deeper understanding of our inner lives, they will undoubtedly reshape our worldview.

Embracing the Mystery of Time

Perhaps one of the most enduring lessons of time travel narratives is the recognition that time itself is a profound mystery—one that touches every aspect of our existence. Our attempts to understand and, perhaps, control time are reflective of our desire to find meaning in our lives. Whether or not we ever unlock the secret to true time travel, the pursuit of this knowledge challenges us to think creatively about our place in the universe and the nature of our own consciousness.

In embracing this mystery, we come to appreciate that our lives are not merely a series of isolated moments but a continuum of experiences—each connected by the threads of memory, emotion, and possibility. The idea that we might one day revisit our past or glimpse our future is as much a metaphor for self-discovery as it is a scientific challenge. It reminds us that the true journey is not about escaping the present but about understanding it more deeply.

The Cultural Impact of Time Travel Stories

Even if literal time travel remains out of reach, the cultural impact of time travel stories continues to shape our collective imagination. From blockbuster films and bestselling novels to philosophical treatises and scientific theories, time travel remains a fertile ground for exploring the human condition. These narratives offer us a way to grapple with questions of fate, free will, and the possibility of redemption.

They inspire us to reflect on our choices and to consider how our past, present, and future are intertwined. They encourage us to value the journey—even with its imperfections—rather than simply seeking a perfect, predetermined outcome. In this way, the enduring appeal of time travel stories is a testament to our deep-seated need to understand and make peace with the passage of time.

Concluding Reflections

As we conclude this exploration of time travel and consciousness, several themes stand out. First, the notion that time is more than a linear progression—it is a dynamic, subjective experience that can be reinterpreted, revisited, and, in some speculative senses, even altered. Trout Theory encapsulates this idea by suggesting that our consciousness is capable of navigating the currents of time much like a fish navigating a river.

Second, while scientific and technological advances continue to push the boundaries of what we can imagine, the realization of true time travel—especially in the mental or physical sense—remains a distant goal. Yet the journey toward that goal is already enriching our understanding of ourselves. The research into brain-computer interfaces, memory transfer, and digital consciousness not only fuels our technological ambitions but also deepens our philosophical inquiries into the nature of identity and existence.

Finally, the cultural resonance of time travel stories lies in their ability to inspire us. They invite us to dream of a future where the mysteries of time are not shrouded in darkness but are illuminated by the light of human ingenuity and curiosity. They remind us that, regardless of whether we can literally travel through time, the true power lies in our ability to reflect on our past, learn from our experiences, and envision a better future.

In sum, Trout Theory remains a compelling blend of metaphor, scientific speculation, and philosophical inquiry. It challenges us to rethink our assumptions about time and to explore the rich interplay between memory, consciousness, and identity. Whether it ultimately proves to be a feasible pathway to mental time travel or remains a vivid metaphor for the human experience, its impact on our collective imagination is undeniable.

As we stand on the threshold of future discoveries, one thing is clear: our quest to understand time is far from over. Every breakthrough in neuroscience, every advance in computing, and every new work of art or literature that grapples with the nature of time brings us one step closer to unraveling one of the most profound mysteries of our existence. Whether we unlock the secrets of time travel physically, mentally, or perhaps in ways we have not yet imagined, the journey itself will continue to shape the very fabric of our culture and our understanding of what it means to be human.

The future of time travel and consciousness is a frontier as vast and unpredictable as time itself. It is a realm where science and art converge, where technology meets philosophy, and where the boundaries of possibility are constantly redrawn. In this ever-shifting landscape, the enduring appeal of time travel stories will continue to inspire us, challenge us, and ultimately, help us to forge a deeper connection with the mysteries of our own lives.

Currents of Redemption

In the grim twilight of the future, when hope was as rare as unpolluted water, Jonas trudged along the cracked concrete streets of a crumbling metropolis. The sky above was a perpetual haze of toxic orange, a constant reminder of the failed experiments and environmental disasters that had brought society to its knees. In this dystopian world, the government ruled with an iron fist, resources were scarce, and every day was a struggle to survive. Yet, amidst this despair, Jonas carried a secret—one that might hold the key to changing the future.

Jonas had grown up in the squalor of the slums, where every life was dictated by harsh realities. The promises of progress had long since withered, replaced by propaganda and repression. But Jonas was different. Ever since he was a child, he had felt time in a peculiar way, as if memories and moments were fluid, like fish darting in a turbulent river. He had always been fascinated by the strange, almost poetic notion that our consciousness could navigate time much like a trout swims upstream.

In a forgotten corner of the city, deep within the derelict basement of a shuttered research facility, Jonas had scavenged bits of old scientific texts and digital archives. Among the yellowed pages and flickering holo-records, he discovered a long-abandoned theory—Trout Theory. It proposed that, under the right conditions, the mind could traverse time without needing to move the body, that the currents of our consciousness might be redirected to revisit past states. Jonas's heart pounded as he read about experiments in lucid dreaming, meditative states, and even hints of quantum phenomena that suggested time wasn't as linear as everyone believed.

Late one rain-soaked night, as the acid drizzle washed the city's grime into forgotten gutters, Jonas sat cross-legged on a cold metal bench in his cramped attic room. Flickering neon signs from the ruined cityscape below cast eerie patterns on his bare walls. With nothing left to lose, he decided to try something radical: to use the principles of Trout Theory to send his consciousness back in time—to his younger self—hoping to alter the events that had led to this dystopian nightmare.

Jonas closed his eyes and focused on the memory of his seventeenth birthday—a day that had set him on a path of despair and rebellion. In that moment, he remembered the bitter taste of loss and the sting of disappointment when his family, struggling under oppressive policies and failing institutions, was torn apart by the regime's cruel measures. It was that day he vowed never to let the future be sealed by fate, even if it meant altering his own past.

He began to meditate deeply, breathing slowly and deliberately, feeling the steady beat of his heart echo through his mind like the pulsing current of a river. He recalled the words of an old philosopher he had once read: *"Time is not a rigid path but a flowing stream—its currents can be navigated if one learns to swim against them."* With every breath, Jonas felt himself sinking deeper into an altered state of consciousness. The boundary between his present and his past began to blur, and images of his childhood emerged in swirling colors behind his closed eyelids.

As the meditation deepened, Jonas felt a peculiar sensation—as though his mind were detaching from his body and drifting along the stream of time. The sensation was at once terrifying and exhilarating. He envisioned his consciousness as a vibrant, shimmering trout, darting upstream against the relentless current. Every thought, every emotion, surged like water through his neural pathways. He focused intently on a singular moment—the moment when he made a decision that had irrevocably altered his life. His vision sharpened until, suddenly, he found himself standing in the memory of his seventeenth birthday.

At first, it felt like a dream. Jonas saw his younger self—a gaunt, disillusioned teenager with eyes too wise for his age—standing in a dimly lit kitchen. The room was sparsely furnished; the air was heavy with the scent of desperation and unspoken sorrow. He could see the familiar faces of his parents, etched with worry and fatigue. The oppressive weight of the future was already palpable, even then, as the seeds of despair were sown with each harsh word and bitter glance exchanged over the dinner table.

Jonas—now present as an observer and a silent guide—watched as his past self hesitated at a crossroads. In that moment, he knew he had the opportunity to change the trajectory of his life. He whispered words of caution and encouragement, though his voice was barely audible, a mere ripple in the fabric of his younger mind. "*Choose differently*," he urged softly, hoping that this echo from the future would be enough to nudge his past self onto a better path.

The memory shifted, and with it came a series of cascading events that Jonas had long since wished to undo. He saw the pain of betrayal by a trusted friend, the spark of rebellious anger that led him into dangerous, self-destructive endeavors, and the gradual erosion of hope that ultimately defined his adult life. All these moments flashed before him like a montage—a turbulent stream of events that he now had the power to influence.

But as the minutes passed, Jonas realized that altering the past was not as simple as merely offering advice. Every action, every word, had consequences that rippled outward, potentially reshaping not only his own fate but also that of countless others. He knew that the world he inhabited was the product of innumerable choices—some made in desperation, some in hope, and many by forces beyond his control. The harsh future he knew so well was not solely his burden; it was shared by a society crushed under the weight of its collective missteps and systemic oppression.

In that moment of clarity, Jonas understood that to change the future, he must change the foundation of his existence—the decisions that led to his despair, the ingrained habits of mistrust and rebellion that had fueled a cycle of destruction. His younger self, still teetering on the edge of a pivotal decision, looked up at him with wide, questioning eyes. Jonas realized that his intervention was a risk—a delicate dance with fate that could either mend the broken threads of his past or unravel them entirely.

Gathering his resolve, Jonas reached out with his consciousness, connecting deeply with the vulnerable, impressionable soul of his younger self. In a series of rapid, vivid mental images, he shared with that young man the insights of the future—a future where hope still flickered, where small acts of kindness and courage could accumulate into monumental change. He conveyed the importance of forging alliances, of resisting the bitterness that had festered over the years, and of finding beauty in moments of fragility and connection.

The memory shimmered as if touched by an unseen current, and Jonas felt a profound shift in the fabric of his past. It was as though a stone had been dropped into a still pond, sending ripples outward that he could already sense altering the contours of time. The scene in the kitchen brightened imperceptibly, the faces of his family softening with a glimmer of renewed hope. His younger self, emboldened by the inexplicable surge of inner strength, resolved to take a different path—one marked by empathy, collaboration, and a cautious optimism that defied the oppressive gloom of the era.

As the sensation of reliving his past began to fade, Jonas found himself slowly returning to the present. He opened his eyes to the familiar dim light of his attic room, the echo of his intervention lingering like a half-remembered dream. Had he truly changed the course of his past, or was it merely a fleeting illusion borne of his desperate hope? He was filled with both cautious optimism and a profound uncertainty that would haunt him for days to come.

In the weeks that followed, Jonas noticed subtle changes in his life. Encounters that once ended in bitterness now carried hints of possibility. The relentless cycle of violence and despair that had defined the dystopian regime began to show signs of disruption. Small acts of rebellion were replaced by acts of solidarity; the voices of the oppressed grew louder, more coordinated, and less fueled by anger and despair.

Jonas's inner journey, guided by the principles of Trout Theory, had not erased the harsh realities of his world. The scars of a dystopian society were still visible in the crumbling infrastructure and the weary eyes of its citizens. Yet, there was a shift—a softening of fate's rigid edges. The past, he realized, was not a prison but a malleable current that could be navigated if one had the courage to swim against it.

As Jonas continued his daily struggle in this transformed yet still broken world, he became a quiet force of change. He organized clandestine gatherings where people shared their dreams and memories, exploring the idea that their collective pasts could be reshaped into a better future. They discussed ancient philosophies and modern scientific theories alike—of how memory and time were interwoven in ways that defied simple explanation. These gatherings, held in secret beneath the ruined city, became incubators for a new kind of rebellion—a rebellion not of violence, but of ideas and hope.

Over time, Jonas and his companions began to see the fruits of their efforts. Policies that had once been accepted as immutable were questioned and reformed. The oppressive government, sensing the stirring of a united populace that no longer saw its past as a chain of inevitable suffering, found itself increasingly challenged by citizens who demanded change. Jonas's intervention in his own past had become a catalyst for a broader metamorphosis—a shift in the collective consciousness that rippled outward like the gentle waves of a long-dormant river awakened by spring thaw.

Yet, the journey was far from simple. With each step forward, Jonas was haunted by the possibility of unintended consequences. The very act of altering the past, even in the realm of memory, carried risks. He would sometimes wake in the night, drenched in sweat, his heart pounding as he wondered if his intervention had inadvertently set off a chain of events that might lead to new forms of suffering. The paradox of having the power to change destiny was both a gift and a burden—a weight that he bore with a mixture of determination and humility.

In moments of quiet reflection, Jonas would revisit that pivotal day in his memory and wonder if, in another timeline, his younger self had made the same choices. He pondered the philosophical implications of free will and destiny: Was he merely an echo of a predetermined future, or had he truly rewritten the narrative of his life? The answers eluded him, but he found solace in the idea that even small acts of courage and kindness could shift the course of history—if only imperceptibly at first, like the gentle ripple of a trout swimming upstream.

As seasons passed and the city slowly stirred from its long slumber of despair, Jonas began to see hope in the eyes of the people. His clandestine gatherings grew into a broader movement, a mosaic of individuals united by the belief that the future was not set in stone. They embraced the idea that while they could not escape the physical reality of their world overnight, they could transform the inner landscape of their collective memory—and, in doing so, chart a new course for the future.

In the heart of the crumbling metropolis, Jonas became a living legend—a man who had dared to swim against the current of time and, in doing so, had ignited a spark of resistance. His journey was a testament to the enduring power of the human spirit, a reminder that even in the bleakest of dystopias, the promise of redemption lay hidden in the depths of our own memories.

One day, as Jonas walked through a rebuilt section of the city—a small park where trees once again reached toward the sky—he encountered a young man with curious eyes. The young man hesitated, then asked in a soft, tentative voice, "Do you believe we can change the future?"

Jonas smiled gently, the lines on his face softening with the wisdom of lived experience. "I believe," he replied, "that the future is as fluid as the river of our memories. We have the power to shape it—not through grand gestures alone, but through every small act of hope, every moment when we choose to see possibility instead of despair."

The young man's eyes lit up, and in that moment, Jonas saw the same glimmer of determination that had once driven his own journey. It was the spark of a new beginning—a future where the lessons of the past could be transformed into a foundation for something better.

As Jonas continued his walk, he reflected on the long, arduous journey that had brought him to this point. His mind drifted back to the turbulent night of his meditation, the moment when he had reached out to his younger self. It had been a leap of faith—a bold experiment in the uncharted waters of consciousness. And though the outcome had been uncertain, that single act of defiance against fate had set in motion a cascade of change that rippled through the hearts of many.

In that moment, Jonas understood that the true essence of time travel—whether physical or mental—was not about escaping the past or controlling the future. It was about embracing the fluidity of our existence, acknowledging that every moment is intertwined with the next, and that the power to change our destiny lies not in defiance of time, but in our willingness to learn from it.

As dusk fell and the city's new lights began to shimmer against the twilight, Jonas felt a deep sense of peace. The dystopian world he had once known was not erased by his actions, but it was slowly being reshaped by the collective will of its people. The harsh future that had once seemed inevitable was now being rewritten, one memory, one choice, one small act of courage at a time.

Jonas paused by a fountain—one of the first installations in the city's revival—and listened to the sound of water cascading over ancient stone. The sound reminded him of the river, of the endless flow of time, and of the resilient trout that had become a symbol of his own journey. In that serene moment, he realized that the future of time travel and consciousness was not a distant, unattainable dream. It was already unfolding in the hearts and minds of those brave enough to believe that change was possible.

The legacy of Trout Theory, with all its scientific speculation, philosophical wonder, and cultural impact, had transcended the realm of mere fiction. It had become a living, breathing force—a reminder that while we may never master time in its entirety, we can still shape our destinies by daring to remember, to reflect, and to act. Jonas knew that the journey ahead would be fraught with challenges, ethical dilemmas, and the ever-present specter of unintended consequences. But he also knew that every ripple in the river of time was a chance to forge a better future.

As the night deepened, Jonas walked on, each step resonating with the hope of a new dawn. The past was not a prison but a reservoir of lessons, and the future was an open horizon waiting to be reimagined. In the quiet determination of his stride lay the promise of redemption—a promise that, even in the darkest dystopia, the human spirit could swim against the current and chart its own course.

And so, beneath the fractured glow of neon and starlight, Jonas's journey continued—a testament to the enduring power of hope, the transformative potential of memory, and the belief that time, however relentless, could be navigated by those brave enough to dream.

Book Description:

Dive into the mesmerizing world of "Trout Theory: A Deep Dive into Consciousness Time Travel"—a groundbreaking journey that blends science, philosophy, and speculative fiction to challenge our very understanding of time. This book invites you to explore the tantalizing possibility that time isn't a rigid, linear path but a flowing river that our consciousness can navigate like a trout swimming against the current.

From ancient philosophical insights and mythological cycles to modern breakthroughs in neuroscience and quantum theory, each chapter unravels a different facet of time travel. Discover how lucid dreaming, meditation, and psychedelics might offer us the ability to "rewind" our minds, allowing us to relive and reshape our past. Delve into thought-provoking discussions on retrocausality, quantum entanglement, and even the futuristic potential of brain-computer interfaces and memory transfer—paving the way toward digital immortality and mind uploading.

The book culminates in a riveting dystopian tale of a man living in a harsh future who unearths the secrets of Trout Theory and finds a way to travel back in time to his younger self. His quest to alter his personal history and, in doing so, prevent a grim global destiny, is a testament to the enduring human spirit and our relentless desire to reclaim hope and rewrite our fate.

Whether you're a science enthusiast, philosophy buff, or lover of speculative fiction, "Trout Theory" offers a fresh perspective on time travel—one that transcends physical boundaries to explore the inner realms of memory, identity, and destiny.

Don't miss out!

Visit the website below and you can sign up to receive emails whenever Walter Grayson publishes a new book. There's no charge and no obligation.

https://books2read.com/r/B-A-CMIKD-INFAG

About the Author

A former computer engineer and amateur philosopher, Grayson has spent his retirement delving into the mysteries of the mind and time. A firm believer that science fiction can predict the future, he writes to bridge the gap between speculation and real-world breakthroughs. His fascination with quantum theory and neurotechnology fuels his explorations into the possibility of reliving one's past. Living in the countryside with his wife and an extensive library, he continues to push the boundaries of what might be possible.

www.ingramcontent.com/pod-product-compliance
Lightning Source LLC
LaVergne TN
LVHW022044200926
848481LV00003B/779